Public Universities and the Public Sphere

Woodruff D. Smith—previous publications:

Consumption and the Making of Respectability, 1600–1800 (2002)
Politics and the Sciences of Culture in Germany, 1840–1920 (1991)
The Ideological Origins of Nazi Imperialism (1986)
European Imperialism in the Nineteenth and Twentieth Centuries (1982)
The German Colonial Empire (1978)

Public Universities and the Public Sphere

Woodruff D. Smith

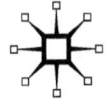

PUBLIC UNIVERSITIES AND THE PUBLIC SPHERE
Copyright © Woodruff D. Smith, 2010.
Softcover reprint of the hardcover 1st edition 2010

All rights reserved.

First published in 2010 by PALGRAVE MACMILLAN® in the
United States - a division of St. Martin's Press LLC, 175 Fifth Avenue,
New York, NY 10010.

Where this book is distributed in the UK, Europe and the rest of the
World, this is by Palgrave Macmillan, a division of Macmillan
Publishers Limited, registered in England, company number 785998,
of Houndmills, Basingstoke, Hampshire RG21 6XS.

Palgrave Macmillan is the global academic imprint of the above
companies and has companies and representatives throughout the world.

Palgrave® and Macmillan® are registered trademarks in the United
States, the United Kingdom, Europe and other countries.

ISBN 978-1-349-29156-4   ISBN 978-0-230-11470-8 (eBook)
DOI 10.1007/978-0-230-11470-8
Library of Congress Cataloging-in-Publication Data

Smith, Woodruff D.
   Public universities and the public sphere / Woodruff D. Smith.
      p. cm.
   ISBN 978–0–230–10878–3 (alk. paper)
   1. Public universities and colleges—United States. 2. Education,
Higher—United States. 3. Social structure—United States. I. Title.
LB2328.62.U6S55 2010
378′.050973—dc22   2010014444

Design by Integra Software Services

First edition: November 2010

10 9 8 7 6 5 4 3 2 1

Contents

Introduction		1
1	The Core Public Sphere: What It Is and Why It Needs Help	5
2	Why Is Public Higher Education in Trouble?	25
3	Building the Core Public Sphere	51
4	The Public Sphere and the Construction of the Modern American University	85
5	Public Universities and the Democratization of the Core Public Sphere	127
6	Occlusion and Its Consequences	149
7	What Should Be Done?	177
Notes		195
Index		223

Introduction

The United States is experiencing a crisis of public discourse. The crisis has many aspects, some of them more widely perceived than others. Everyone who has followed the news for the past decade is aware of the policies proclaimed by the federal executive branch that give agencies of the national government the ability to declare practically any kind of information secret. This situation appears to be a reflection of a general tendency throughout a large part of the world. It represents an obvious threat to the intelligent discussion of issues affecting the public interest, and possibly to the traditional concept of "the public interest" itself. The current administration has begun to reverse such policies, but how far the reversal will go in practice remains to be seen.[1] Most readers of newspapers are also aware of the connected set of issues that involve the ownership and practices of the mass media. Are too many media in too few hands? Do the owners have too much influence over the political positions taken by the periodicals and television networks they control? Is open, rational public discourse being stifled and replaced by dumbed-down gibberish masking the selfish interests of media owners and unscrupulous politicians, and the agendas of the groups they support? Al Gore, among a host of others, warns us that it is.[2] Then there is the matter of the internet. Is it the most likely source of a satisfactory response to the threat described by Gore, or will it destroy the structures on which informed discussion depends—as defenders of traditional newspapers and opinion magazines predict?[3]

These are important matters. There are, however, other aspects of the crisis of public discourse that have, at least until recently, received less attention. One is the threat to the existence and independence of the "quality" news and opinion media that is posed by, among other things, unstable or declining subscription rates. The specific problems of the quality media are usually folded into discussions of the plight of the large daily newspapers that have recently faced bankruptcy and closure. This is not unreasonable, but it leads to an emphasis on the danger to the availability of accurate news at the expense of the equally significant danger to critical debate over issues. Another threat, although it is coming to be more widely understood than it has in the recent past, is seldom seen in the context of public discourse at all: the possibility that the funding of state universities, already in decline for many years, will diminish even more sharply, putting an end to the growth of public higher education that has been a central feature of American society for the past century. Both of these threats are vitally important. They are also intimately related to each other. We will see in the following chapters that recognizing the relationship between them is essential to resolving satisfactorily the overall crisis of public discourse in the United States.

The key to this recognition lies in perceiving that there exists in the United States—as in most other countries—an entity that we will call here the "core public sphere." It is composed of the quality media, the universities, and an array of other structures such as professional and research organizations, as well as the practices associated with them. The function of the core public sphere is to formulate, develop, and openly debate issues of importance to the public, and to do so intelligently and with reference to the best information available. In order to perform this function, the participants in the sphere must have ready access to a full range of presentations of relevant ideas, the ability to criticize any such presentations that they regard as incorrect or inadequate, and the freedom to advance their own ideas. These are not just passive capacities that can be described in terms of legal rights, although constitutional guarantees of free speech are obviously essential to their exercise. The prerequisites for public discourse must be embodied in institutions that permit their effective realization and in behavior patterns that conform to acceptable standards of logic and civility. They must be regularly put into practice by people who acknowledge a responsibility to participate in public discourse and to educate themselves so that they can do so effectively. Participants in the core public sphere are not only the people who produce materials for publication or transmission, but also—and equally importantly—the people who read, view, as well as think about, discuss, and otherwise respond to what they read or hear: what might be called the "discursive public." In a wide variety of ways, higher education is central to practically every aspect of the core public sphere.

The core public sphere is neither an abstraction nor something that necessarily appears in modern societies simply as a consequence of modernity. The United States did not possess a core public sphere of its own until the second half of the nineteenth century, when one was deliberately constructed. China does not possess one now. Core public spheres do not always perform their functions very well. They often display serious weaknesses, the most common of which are tendencies toward elitism and intellectual involution. Nevertheless, societies that do not support effective core public spheres can pay a heavy price—as the United States did before the Civil War and as China does today. Considering the role of the United States in the world and the weakness of public discourse in the only other country that could conceivably play that role in the near future, any serious degradation of the core public sphere in the United States could have serious global consequences.

The chapters that follow will argue that the American core public sphere is facing a serious crisis, a significant aspect of the broader crisis of public discourse. Public higher education, also facing substantial problems, would be in a position both to help resolve part of the crisis of public discourse and to solve some of its own difficulties if it acknowledged its actual roles in the core public sphere and reoriented some of its practices around those roles. A major part of the argument revolves around a historical examination of the connection between higher education and the core public sphere in the United States. The first chapter describes the American core public sphere and the nature of the problems it currently faces. After that, the book considers some of the principal problems of public universities, weaknesses in their self-presentation, and some of the undesirable effects of the academic ideologies that purport to explain, but do not adequately describe, what public universities do. The historical segment begins by outlining the close relationship between higher education and the core public sphere since just before the time of the Civil War. It then discusses the particular contribution of the state-supported colleges and universities in democratizing the public sphere and the vital significance of this function. A chapter describes how this function, together with the nature and implications of the connection between higher education and the public sphere in general, came to be occluded in public discourse in the twentieth century, even as the connection became increasingly central in practice to defining American society. The final chapter returns to the present and suggests what public universities should do to resolve satisfactorily their own crisis, some of the problems of the core public sphere, and the larger crisis of American public discourse.

CHAPTER 1

The Core Public Sphere: What It Is and Why It Needs Help

The German philosopher Jürgen Habermas introduced the idea of "public sphere" into academic discourse in 1962. It came to be widely employed in scholarship and intellectual conversation in the English-speaking part of the world only in the 1980s, after Habermas had achieved international renown for other work, but it has become enormously popular since then. Actually, the word Habermas used was *Öffentlichkeit,* for which no ready English equivalents exist that convey Habermas's sense. "Public sphere" does as well as anything else.[1]

To Habermas, the public sphere was a vital structural component of bourgeois society in the course of its evolution, particularly in eighteenth-century Europe. It appeared at the intersection of several of the crucial developments that created the modern world: the growth of national and international market economies, with their need for expanded sites of communication and exchange and for a clear definition of individual property; the construction of the modern state, institutionally abstracted from traditional systems of status and custom; and the redefinition of the nuclear family as the primary locus of the "private," the space closed off from legitimate interference by the state and the place where individual personality emerged. One of the results of the confluence of these developments was "civil society," classically described by John Locke in the late seventeenth century. Although sometimes the terms "civil society" and "public sphere" are conflated, Habermas treats them as either separate, although related, concepts, or else nested historical realities: the public sphere as a distinct part of civil society.[2] Either way, the public sphere supplements civil society by defining a space separate from the spaces of government and politics where broadly based, free discussions take place that lead to the formation of public opinion, which among other things guides and legitimates legislation.[3]

According to Habermas, the structures that constituted the original public sphere—coffeehouses, clubs, newspapers—first appeared in the late seventeenth century in England, flourished there in the next century, and were imitated in other countries in the late eighteenth and early nineteenth centuries. He goes on to claim that, to all intents and purposes, the public sphere passed out of existence in the course of the nineteenth century because of the triumph of "mass" society dominated by organized capitalism, "mass" media of low intellectual content, and "mass" political parties.[4] This last point leads Habermas to formulate a goal. He argues that the *concept* of the public sphere as a necessary condition for an effective civil society is just as valid now as it ever was and that something like the historical public sphere needs to be reconstructed.

To scholars and commentators who employ the model of the public sphere in order to analyze the contemporary world, the idea that at some point it ceased to exist presents an obvious problem. For the concept to be useful, it must correspond in some substantial way to a current reality, not just to a goal. Like most of the people who use the term, I do not agree that the public sphere essentially disappeared in the course of the nineteenth century.[5] I think that instead it was consciously modified and adapted to new conditions. In various countries and at various times, public spheres were constructed to meet perceptions of changing need. Their construction took place as one aspect of a set of broader developments, including many of the ones that Habermas blames for destroying the classical eighteenth-century public sphere: the appearance of a "mass press," popular politics, and the consumer economy in particular. Newspapers and magazines that emphasized politics, public affairs, and high culture acted as central institutions of the constructed public spheres in Britain, France, and most other Western European countries. These media embedded themselves in networks of writers, editors, party politicians, officials, and scholars who contributed to them and connected themselves with voluntary associations that supported extensions of their conversations. The growth of the popular press and of advertising toward the middle of the nineteenth century vastly expanded the range of journalism and public opinion, but it did not drive the kind of public spheres developed earlier in the century out of business. Rather, these public spheres remained (and remain to this day) significant features of public life. No distinctive, commonly accepted categorical name has ever been created for them. We will call them the "core public spheres" of the countries or regions where they exist. If we were focusing just on the nineteenth century, we would almost be justified in using a name such as "elite public spheres." They were certainly elitist in many respects, but even in the nineteenth century they had a substantial audience among the middle classes—the people who became the principal consumers of periodical literature and the ultimate arbiters of opinion and politics. As we will see, in the twentieth century they became even more democratic, although never completely so.

The Core Public Sphere in the United States

The American core public sphere was deliberately constructed in the late nineteenth and early twentieth centuries. The Americans who built it did not start from scratch, but rather used European models and precedents with which they were already very familiar because Americans had up to that time essentially shared in an Atlantic public sphere, the center of which was Great Britain. The American core public sphere has continued to exist, with modifications, down to the present. It has continued to maintain a strong transatlantic character, but in the 1870s it renounced its dependent status. We will discuss the history of the American core public sphere in Chapter 3. For the moment, we will focus on its elements.

The principal activity or function of the core public sphere can be envisioned as the one that Habermas emphasizes: the open, critical discussion of the whole range of matters of concern to society, with the conscious understanding that both discussing and deciding what is of concern, what is important, are primarily tasks of people qualified to do so by virtue of intelligence, education, occupation, taste, and especially interest. This activity could easily be comprehended in an elitist way, as it certainly was in the late nineteenth century. A number of factors in American culture and society, however, combined in subsequent years to democratize the core public sphere—the most important of which was, as we shall see, the development of public higher education. This principal activity takes place in particular *structures* and according to certain conventional models of *practice*.

The *structures* include what might be called the "quality" media. Some of these, especially weekly, monthly, and quarterly periodicals such as the *Nation, Atlantic Monthly,* and the now long-defunct *North American Review,* were envisioned in the nineteenth century as the primary sites of intelligent public discussion and the principal organs of the new public sphere. They were intended to support a conversation among educated contributors and readers from which ideas and opinions would diffuse through the popular press to the country as a whole.[6] Parties to the conversation would include not just journalists and writers, but also officials, politicians, and university professors. Periodicals of this sort (including, obviously, some of the very same magazines) continue to perform a central role in contemporary public discourse, despite the decline in their readership relative to the size of the population. So do certain newspapers, preeminently the *New York Times,* whose editors have for over a century deliberately molded it as the chief daily news medium of the American core public sphere. Other media structures were created later: particular departments of general newspapers (editorials, op-ed pages, Sunday "ideas" supplements); equivalent sections of news weeklies; and public affairs television programs (especially on Public Broadcasting). To these must be added a range of other familiar structures: think tanks, foundations, policy centers, scientific research institutes—and of course, universities. For the moment, however, we will focus mainly on the media element.

The core public sphere is also a space for the conduct of certain *practices*—not the exclusive site, but the principal area within which they are applied to articulating and debating issues of importance to the nation. The identities of these practices have changed somewhat over time. In the late nineteenth century, for example, what we are calling the core public sphere was seen as one of main spaces for practicing *science.* Science subsequently became fully professionalized and the principal locations of its practice moved conceptually to specialized institutions (mainly to universities, as universities came to be thought of as lying somewhat outside the arenas of public discourse.) Nevertheless, science has continued to provide the most highly valued model for thought and for discursive practice within the culture of the core public sphere. We presume that the ideal perspective on most matters within the scope of the public sphere is essentially a scientific one, requiring objectivity, detachment, disinterest, and an insistence on rational analysis—or if that is not possible, then at least a frank admission of partiality and interest, a commitment to making arguments in a rational manner with convincing evidence, and a willingness to tolerate the expression of divergent views. (One of the assumptions that we commonly make—correctly or not—is that the more widely read and accessed particular media are, the less they are capable of sustaining the kind of scientific discursive practice that is expected in quality journals and newspapers and that is supposed to be practiced in universities and research institutes.) By the twentieth century, the social and behavioral sciences took on a particularly important function as the presumed "objective" arbiters on matters of public interest, operating from an institutional position notionally (although seldom actually) outside the arena of debate.[7]

In the nineteenth century, before social science gained the starring role of model for the practices of the public sphere, *history* served in that capacity. History identified parallels between past and present issues. It was supposed to deploy systematic reasoning and carefully assessed evidence to advance interpretations of past events that were relevant to the contemporary world. History made it possible to reconcile a discourse of character and morality with a practice of objective science. The framework of history as a principal practice of the public sphere essentially defined the subject that graduate history programs were established to propound in American universities in the late nineteenth century. Although history lost its position of primacy to the social sciences, it still retains much of its significance as a mode of analyzing and arguing in public discourse.[8]

More broadly, the core public sphere, like Habermas's classic eighteenth-century public sphere, is a space for the practice of *criticism*. Rational, disciplined criticism, brought to bear on matters of importance, has been its most enduring characteristic. The practice of criticism naturally creates conflict and tensions. One of the many reasons that universities tend to underplay their roles in the core public sphere is that criticism is so central to the latter. University leaders perceive that those who make funding decisions for higher education are frequently very sensitive to criticism and tend to see it as an activity that should be restricted to arenas not supported with public funds.

Many of the practices located conventionally within the core public sphere are associated with *professionalism*—which is a manifestation, in the normal behaviors and attitudes required from members of particular occupational groups, of the objective and scientific practices just mentioned, and also an institutionalization of standards that give such practices ethical standing.[9] The people who created the core public sphere expected that members of the traditional gentlemanly educated professions (law, medicine, the clergy) would supply many of the active participants in public discourse. Two occupations directly connected to the operation of the public sphere were deliberately professionalized: journalism and college or university teaching. Both occupations had existed in the United States for a long time, but not as recognized professions that required specialized training and that enforced particular forms of practice and ethics. In the late nineteenth century, journalists and reformers consciously attempted to create a journalistic profession that operated in the public interest. The importation of the German concept of the doctorate based on research into the United States starting in the 1870s was similarly part of an immensely successful attempt to create a professoriate, not only to staff the new and reformed universities but also, as we shall see, to provide a significant professional addition to the personnel of the core public sphere.[10]

To summarize: the particular arena of American public discourse that we will be concerned with in this book is the *core public sphere*. It has some of the same functions and characteristics that Habermas ascribes to the classic British public sphere of the eighteenth century: it is supposed to support an open, critical, informed, and intelligent discussion of matters of importance to the country and the world; to make access to the discussion available to people who want to participate, whether actively (e.g., as contributors to publications, on the understanding that what they contribute is open to public criticism) or more passively, as readers and perhaps private or local discussants of issues; and to produce a rational consensus on particular topics that can be presented to a wider body of citizens for acceptance as "public opinion" or else clearly articulate alternative interpretations of topics and, if appropriate, the policy implications of those interpretations.

Unlike the public sphere described by Habermas, the modern core public sphere performs these functions through a complex group of institutions that are expected to interact closely with political and economic structures, among other things in order to obtain information and to ensure that public discussion affects policy. They are also, however, supposed to maintain a separation from these structures so as to retain a critical perspective. These characteristics create familiar problems that are themselves continuing subjects of debate in the public sphere. For example, how do you balance the need of, say, the quality newspapers for revenue from commercial enterprises and for information from government with the need of the public sphere to maintain critical distance and with the desire of governments to shape the news? This is one of the tasks that professionalism is supposed to accomplish, through the ethical framework of journalism. Also, although we expect newspapers and newsmagazines to be "objective" and "neutral" (except on the editorial pages, and even there a degree of balance is demanded), we also assume that "opinion" media (the *Nation, National Review*, etc.) will have particular political and ideological tendencies—which are to be clearly expressed rather than dissembled. The *range* of points of view expressed across the public sphere as a whole, representing in many cases the competing interests of economic and political groups, will supposedly compensate for the inherent inability of the modern core public sphere to separate itself from the institutions of state and economy in the way that Habermas describes the eighteenth-century public sphere as doing. Leaving aside the likelihood that, in reality, the eighteenth-century public sphere was not as detached from political and economic interests as Habermas says, the fact remains that the core public sphere of the contemporary United States, like its counterparts in most other countries, performs a function similar to the one Habermas describes, but does so through structures and cultural patterns that are in some ways quite different. They are literally more *modern*—both the structures and the behavior patterns that are supposed to be characteristic of the core public sphere are quintessential elements of modernity.[11]

The previous paragraph referred to the "supposed" or "expected" functions of the core public sphere. This usage implies that the core public sphere is something that is recognized as both an ideal form and a structural and cultural reality that can be measured against the form. This is true, with qualifications. There is no accepted name in common usage for what I have called the "core public sphere," which, as we will see, has been a source of difficulty for a long time. In order to name it, I have adopted and modified Habermas's terminology. This suggests one of the points that will be made later: although the existence and importance of the core public sphere are recognized in the United States, the recognition is informal and is partly obscured by, among other things, the conceptual framework within which the public sphere is understood.

"Core" suggests the simultaneous existence of a periphery. The people who created the American core public sphere in the late nineteenth century thought—or at least hoped—that public discourse in the United States would actually arrange itself in a core-periphery format. The quality periodicals, the conversations carried on in them, and various ancillary elements (such as universities) would act as the center, while the rest of the press, thought of mainly as popular periodicals, would cluster around them. The center would provide the brain to the body of the discursive public; the rest of the media would supply revenue and careers and the principal means for disseminating the results of the conversations at the center.[12] Things did not quite work out that way. The structure of the American media became immensely complex, and the quality periodicals did not develop as hegemonic centers of public discussion. Nevertheless, for more than 100 years, there has been a general understanding among people who regard themselves as opinion leaders that there is a kind of structural and institutional center for public discussion and that this center is supposed to operate in a particular way.

One problem that arises with this view is deciding where the center leaves off and the rest of public discourse (the "broad public sphere," perhaps) begins. This is related to the question of which structures and which participants are to be held more or less strictly to the behavioral norms associated with the core public sphere. Thus, for example, the *New York Times* and the *Wall Street Journal* and their staffs, the *Atlantic* and *Harper's*, the *Nation* and the *National Review*, the Brookings Institution, the news and information programs of the Public Broadcasting System, Harvard University—all of these are clearly examples of central, prestigious elements of legitimate public discourse in the United States and thus fixtures of the core public sphere. The *National Enquirer*, Howard Stern, and diploma mills are not. In between, however, is an enormous range of institutions, media, and people associated with them. Some may be assigned informally to the center or the periphery of public discourse depending on self-presentation and the judgment of the person doing the assigning. Some may contain certain elements deliberately dedicated to serious public discourse and other items devoted to different functions: for example, the editorial as opposed to the comics pages of an ordinary newspaper (although there is always *Doonesbury*). Nevertheless, however vaguely its boundaries may be defined, we generally assume that there exists an institutional and professional center for significant public discourse. We can think of it as being situated at one end of a continuum, with the quality and opinion periodicals (including those online, like *Salon.com*) at one end and the *National Enquirer* and crazy blogs at the other, and we can—if we wish—distribute the other media organs where they seem to fit: *Newsweek* and *Time* not far from the "core" end, the *New York Post* (in its contemporary incarnation) not far from the other end. (There is irony in this. At the time of the core public sphere's creation in the nineteenth century, the *Post* essentially defined the elite, intellectual end of the continuum among daily newspapers. Only gradually did the *New York Times* replace it.)[13] Or we can simply accept the perception that, I believe, most readers of serious periodicals share: there is a group of media that we expect to take the lead in conducting significant public discourse, the boundary of the group is fuzzy due in part to the multiform character of individual organs and their owners' desire for as wide an audience as possible, and the media in the group bear the largest share of the responsibility in our society to supply well-researched information and reasoned interpretations of it. When we talk about the contemporary problems of the media, of newspapers, of public discourse, it is primarily this group that we have in mind. It has been the central element of the core public sphere since the nineteenth century, although it is not the only one. We will look later at some of the others, particularly those connected with universities.

Why the Core Public Sphere Is in Trouble

As we saw in the introduction, the most obvious and heavily publicized difficulties of the core public sphere are financial. Many of the principal media are experiencing a crisis of solvency. But before we address that problem, we have to consider something that has been evident for some time: the American public sphere is also facing a *crisis of performance.*

Many examples could be cited to illustrate what is meant by a "crisis of performance." One, however, will suffice: the response of the American media and other institutions that are supposed to examine national issues critically to the events of September 2001, and also to those that led in 2003 to the invasion of Iraq. The latter was a failure of the nation in practically every respect (except perhaps in a short-term military context that dissolved almost immediately.) The Bush administration certainly should be held responsible for making and adhering to unbelievably bad policies and for displaying monumentally bad faith. But public discourse also failed. The administration contributed heavily to this fiasco through its practices of information management, but it had a lot of help.

The response of American public discourse, and particularly of the core public sphere, to the events of September 2001 showed not only poverty of imagination and inability to put events into intelligible perspective, but serious deficiencies in structure and practice. The obvious need for thorough discussion of the complexities of a world that suddenly showed that it did not conform to conventional pictures Americans—even highly educated Americans—had adopted of it, was met by a muted response from universities and from the prestigious media. The narrow and puerile character of public conversation could, for a while, be explained by panic, but it continued long after fear had sanctioned ill-considered military actions culminating in the invasion of Iraq, the results of which were entirely predictable. A properly functioning system of public discussion should have foretold the consequences much more convincingly than the existing structure did. Why didn't it? In part, because the audience for public discussion was content with (indeed, insisted on) inanity and sloppy thinking in place of informed debate, which displayed among other things a weakness in its members' education and preparation for dealing with complicated public issues. In part also, because the institutions that frame and support public discussion—especially the media and the universities—perceived severe limits to their ability to withstand pressure and to promote consideration of ideas that lay outside the range of conventional thinking. They failed, therefore, to fulfill their self-proclaimed responsibilities. Although events have subsequently forced a change of public perspective on the Iraq invasion and a more comprehensive retrospective on the process that led up to it, I see little evidence that much has altered with regard to the deficiencies in the core public sphere that 2001 revealed.[14] The same thing would, under similar circumstances, happen again. Simply changing administrations, simply listening to media news figures bemoan their failures in 2001–3, does not suffice.

One point should be emphasized: it was not just the structures of the core public sphere that failed in 2001–3; the regular *audience* of the quality media, the rank-and-file participants in serious public discourse, also failed. This portion of the public sphere is more difficult to identify than the professional personnel of the major institutions. Except for university students, it is not usually signified by organizational affiliation. In general, though, it consists of people who read or watch the prestige media and who regularly discuss what it is in them, who regularly read (and sometimes respond to) the editorial pages of newspapers. They constitute an informal and self-selected but generally recognized group. (Think of the principal audience toward whom Public Broadcasting targets its programs—and its fund-raising appeals.) They typically see themselves as opinion leaders and often are seen in this way by others. In some general respects—education, interests, tastes, self-image, perhaps class background—the majority of such people have much in common with one another, but as a group they are far from uniform or exclusive. The growth of the blogging community has made the traditional modes of identifying them even less clear. Entry into their ranks is largely a result of a conscious decision to take part. For the most part, these people have expectations about themselves that conform to an understood (if unenforceable) set of social responsibilities. They are supposed to reason intelligently about public affairs and to insist on a high level of criticism and analysis from the professionals at the core of public discourse. A great many of them did not do these things in 2001 and in the succeeding years. This circumstance has to be taken into account in any consideration of what is wrong with the core public sphere.

The current *financial crisis* of the quality media is very real and very important, but problems of revenue are nothing new in the core public sphere. This is obvious from looking at the institutional fortunes of the monthly and weekly magazines that have traditionally served as its center. *Harper's* suffered substantial losses in readership in the 1970s and almost went out of business in 1980. It was saved only by foundation support, a rallying of the readers who remained, and a substantial reorganization.[15] It has ceased to publish as wide an array of types of article as it once did—ostensibly because other, more specialized magazines are available. The *Atlantic* underwent several reconstructions in format and audience in the twentieth century (shedding much of its overtly elitist orientation in the process.) It also experienced fairly substantial rises and falls in readership. By the mid-1990s, it appeared to have recovered from a long-term decline and regained a circulation of about half a million, but that has fallen by about 20 percent since then.[16] The *Nation*, the *New Republic*, and most of the other long-standing "opinion" periodicals have displayed similar trajectories. In most cases, their survival has depended on infusions of money in the form of new ownership willing to operate at the margin of profitability or at a loss. The *Nation* got a new lease on life when its circulation shot up because of liberal response to the Bush administration's policies in 2001–2; the *New Republic* has approached collapse since that time in part because it was perceived as supporting the Bush administration and generally as having moved to the right. The *National Review* held its own through the mid-2000s, but its circulation had fallen by nearly 50 percent from its peak in 1994.[17]

The quality periodicals have never had an easy ride. Many have come and gone since the middle of the nineteenth century, and the circulations of the ones that survived have typically fluctuated as those of the magazines cited previously have. They have frequently required what amounted to subsidies in order to stay in business.[18] But generally speaking, the problems of quality periodicals were individual; when one experienced falling circulation, some of its rivals saw gains, except in periods of general economic depression such as the early 1930s. The overall decline in the circulation of the qualities since the 1970s is something unusual—and disturbing, if one thinks that what they provide is vitally important and not really offered elsewhere. It should be noted that the current decline set in long before the challenge from the new internet media became significant. The contemporary online equivalents of the quality periodical press have displayed similar weaknesses in subscriber demand.[19]

The problems of the quality magazines have recently been overshadowed by the impending disaster of many of the major daily newspapers. In the past few years, demand for daily newspaper subscriptions has fallen precipitously, which has caused a near-collapse of advertising revenues.[20] The *New York Times* recently had to be bailed out of insolvency by a huge subsidy; its parent company, which also owns the *Boston Globe,* recently threatened to shut down the *Globe* unless something staunched the *Globe*'s losses. (The "something" was union agreement to major wage and benefit cuts.) The *Los Angeles Times* and the *Chicago Tribune* are bankrupt. The Denver *Rocky Mountain News* is out of business. The list is large—so large that some observers have called for the government to provide subventions to failing newspapers.[21] (Note that while some of the newspapers that are in trouble or have folded are—or were—located in cities where there is another major newspaper of at least equivalent quality, this is not true of the *Boston Globe* or the *Los Angeles Times.* And it could be reasonably argued that a city or region with only one high-quality newspaper is not being well served from the standpoint of public discourse.)

The proximate cause of the problems of the large daily newspapers is, these days, usually held to be the internet. In the 1990s, television news was the favored culprit, although the problem was less acute. Businesses have been moving their advertising away from the print media; an increasing number of people get their news entirely from internet sources or from television. The two principal sources of newspaper revenue, advertising income and subscriptions, have fallen accordingly. What the current situation has exposed is the dependence of much of the American public sphere on its ability to facilitate commercial marketing.[22] The problems created by this dependence have been pointed out for years, but the focus has mainly been on the reportorial or editorial bias that might, and not infrequently does, result.[23] Only recently has the possibility that it will lead to the dissolution of print newspapers as a genre come to the forefront.

Part of the problem stems from poor decisions by particular newspapers, including their failure to find ways of adapting their operations to the internet. But if that were all there were to it, newspapers could just make new decisions (as many are trying to do) and move their operations online.[24] The ones that managed to make the change efficiently would survive, and the others would not. Although it would be a great pity if some of the comprehensive newspapers were to fail entirely, there are many ways in which their functions could be taken over by other institutions or by revised versions of themselves. As we will discuss later, the internet has created the possibility of integrating the audience of the core public sphere much more closely with its institutional structures than ever before, and also (through blogging, among other things) counteracting some of the core's inherent elitism. The current crisis of the news and opinion media could therefore be the occasion for a very positive reworking of public discourse. Al Gore certainly thinks so.[25]

Unfortunately, this may not be easy, partly because of economic factors and partly because of other factors that lie behind them. For example, when advertising leaves the newspapers for the internet, there is no particular reason that it has to go to the online equivalents of newspapers. There are plenty of other sites that sell advertising space and that attract large audiences. Although some online news-sites are doing better than others (especially ones that essentially retransmit the news they get for free from the regular media), most are still nonprofits. It appears that the newspapers that are planning to recover their finances by concentrating on the internet intend to rely on subscriptions; the jury is still out about whether or not this will work.[26] The fact is that newspaper subscription rates have not been expanding much for most newspapers for many years—from long before the availability of internet news.

Moreover, using electronic means of distribution and communication in place of print and paper, though it may cut some costs, does not eliminate many of the most important expenses borne by the print media—most notably, the very high costs of professional reporting, not only at the national and international levels but also at the local level. There are also costs associated with research for articles in opinion media. *Someone* has to pay for these things. If advertising were to move to the internet specifically to support electronic media, the traditional commercial arrangement would be replicated. There would still be the problem of commercial interference with what is discussed and reported, but at least that is a familiar difficulty for which familiar (if not always wholly effective) remedies exist. But there is no guarantee that this will happen. As we will see later, these and other factors will almost certainly force us to reconsider the whole nature of the core public sphere (which is one of the reasons that it is important to name and recognize it). In a moment we will ask the crucial question: is the core public sphere actually necessary? Later in the book, we will ask whether, if it is necessary, the forms that it takes must mirror those that it has taken in the past. But regardless of the shape that the core public sphere takes in the future, it seems unlikely that the newspaper part of it can be funded as it has been in the past.

In this regard, it is helpful to think about television. One of the reasons that a smaller proportion of the American population reads newspapers now than in the past is that the majority of people get their news from television.[27] This has been the case for quite some time. Leaving aside criticisms of television as a "passive" medium that does not stimulate viewers' critical faculties but rather stirs their fears[28] (reading newspapers can be fairly passive and scary as well, if you let it be so), it is certainly true that a standard television news show covers far fewer stories, in less depth (or at least with fewer words), and with less opportunity for thoughtful editorial comment than a reasonably good daily newspaper. It is not the case that news and commentary cannot be presented well on television; rather, it was discovered in the early days of television (and before that, radio) broadcasting that if they were, they did not pay for themselves by attracting enough of an audience to bring sufficient advertising revenues. The kind of news coverage that we have now represents a compromise between, on the one hand, considerations of profitability and attractiveness to the largest possible audience, and on the other, an effort to meet at least the minimum standards of the core public sphere and the portion of the audience that accepts and demands those standards. It is probable that if television news were to become truly dumbed down, it could make more money (witness the fact that, at the national level, the news programs least concerned with the standards of the core public sphere appear to be the most successful at attaining a large audience). But even with the best of intentions, the demands of the market seem to make the present compromise necessary. Networks and local stations do offer sophisticated programming and significant commentary, but these do not generally contribute to profitability. They are, in essence, *subsidized* by the rest of the broadcasters' offerings.[29]

It is not surprising, then, that (with the wholehearted blessing of most of the commercial networks) public television should have taken over the function of providing the majority of the in-depth, significant coverage of news and debates over issues, together with the other programming that it puts on the air. The Public Broadcasting System (PBS) is paid for by commercial and noncommercial sponsors (the former of whom get a limited amount of advertising value out of it, among a targeted audience), by subscribers (who are entirely voluntary, since access is available without subscribing), and by *subsidies* from the federal government.[30] Without the subsidies, public television would be unable to do what it does—especially in the area of detailed coverage of news and issues. The tendency over time for the federal subsidy to be reduced has created something close to a crisis in public broadcasting. The threat of further reductions unless political coverage was more "balanced" (from the standpoint of the party in possession of the White House or a congressional majority) has demonstrated on several occasions the downside of having an important part of the core public sphere dependent on government subsidies.[31] Nevertheless, it is clear that unless we want to do without PBS (which I, for one, would not), we have to accept the idea of subsidies.

The dependence of PBS on government subsidies is actually even wider and more complex. Not only are its more spectacular domestically produced programs partly paid for by subsidies from agencies themselves funded by American state and federal governments, but a substantial part of its most-watched programming is actually produced by state-supported *foreign* broadcasters like the BBC. The latter have been attempting to make up for reductions in their own subsidies over the past two or three decades by increasing their sales to American public television. And in the cases of non-state-supported overseas television productions that appear on PBS (e.g., ones put out by British Independent Television), what is happening is that quality foreign programs are being paid for in part by subsidies (public and private) originating in this country.[32]

This may seem to be excessively complicated, but actually it is an example of one of the more positive aspects of contemporary globalization. In essence, the system works on the principle of comparative advantage, expands the size and transparency of markets, and creates economies of scale, to the benefit of practically everyone involved. But it works only because it involves the active participation of states as well as of nongovernmental organizations, businesses, and individual subscribers on an international basis. It is an example to be kept in mind when we return in the final chapter to the question of how the core public sphere could be preserved or altered to make it work better. As we will see in Chapter 3, an important part of the core public sphere in the United States has always been subsidized, one way or another. Subsidy will clearly have to be a part of whatever is done now—the price of effective public discourse.

Of course, this presumes that something actually should be done, that the core public sphere is sufficiently useful that steps should be taken to preserve and improve it. If it is not performing up to a reasonable standard, if it is experiencing a crisis of performance, could that mean that it is intrinsically incapable of functioning properly? As we will see in later chapters, the classic American core public sphere has suffered from a number of deficiencies for a long time. Since its construction in the nineteenth century it has had an elitist character, which was reduced substantially in the twentieth (by, among other things, the growth of public higher education) but never completely eradicated. The separation between the core and the broader, more general elements of the public sphere is one of the reasons that the former has not been as effective as it could have been anytime in the past century. Problems of performance are also nothing new. The American core public sphere has regularly let the country down in wartime, failing to provide sufficient critical perspective and debate when it was most needed and, usually in the name of patriotism, truckling under to the government and to majority opinion and agreeing to the silencing of critics. This was true of both world wars and, for a while, the Vietnam War. The media of the core public sphere also took their time about responding to McCarthyite red-baiting in the 1950s, as they had to Wilsonian red-baiting in 1918–20.[33] The remarkable shift of American public opinion in 2005–6 against the Bush administration was clearly not due to a consensus among the quality magazines and newspapers, because there was no such consensus.[34] The core public sphere took until the 1950s at the earliest to develop a consensus that action was needed immediately to end racial segregation.

On the other hand, in all the cases just noted, the core public sphere sooner or later did perform its function. In the late 1960s, it was the eventual emergence of a consensus in the core media that led to the widespread agreement among "responsible" people that direct American involvement in the Vietnam War had to be ended (the sort of thing that did not happen with regard to Iraq until a popular consensus announced itself in polls, the general media, and the ballot box in 2006).[35] Even in recent years, it has not infrequently taken the lead. For example, the core public sphere, pretty much across its entire ideological spectrum except for a divided right, some time ago developed a consensus in support of equal rights for gay and lesbian citizens. It did so well in advance of general public opinion. Moreover, the fact that in the past couple of years the broad discursive public has been performing some of the functions that the core public sphere was originally supposed to undertake probably suggests not that the latter is unnecessary, but that we need to reassess the modes of its operation and the way the core elements relate to the rest of the public. Something interesting is going on that we need to understand.

In any event, as we will discuss in later chapters, it is very difficult to imagine a modern liberal democracy operating without at least a reasonably effective center of informed public discourse. It is entirely possible to imagine a modern state that was not a liberal democracy and that therefore might not need (and whose leaders certainly would not want) a core public sphere. There are several such states currently in existence, and the past century was replete with experiments with alternatives. The most spectacular examples of the latter (Soviet Communism, Nazism, Fascism) all failed, and they failed not just because they did not measure up to a liberal standard but because they did not work, even in terms of their own criteria for success. With regard to the current situation, it seems to me that one of the most serious factors impeding China's full development as an effective modern state is precisely the absence of a recognized, legally permitted public sphere, which among other things means that major problems that arise in Chinese social, economic, and political life are almost never discussed and thought through before they get out of hand.

At least tentatively, let us assume that something like a core public sphere is necessary in the United States and that its current crisis of performance justifies not its abandonment but its improvement. The historical discussion in the central part of this book will provide stronger justification for both parts of the assumption. Let us turn to possible explanations for some (far from all) of the current problems of the core public sphere, especially ones that have implications for higher education.

Certain explanations are heavily featured in contemporary debates over the state of journalism: the commercial character of most of the media, which may cause them to be influenced by what their owners and their advertisers want them to say and by anticipation of what the audience wants to hear or will tolerate; practices of information control on the part of the government, which have been extremely overt in the 2000s but have never been absent; and changes in the professional standards of magazines and newspapers, often ascribed by critics to the advent of the "new journalism" (which is not so new anymore) or to declining standards among the readership.[36] Some of these factors are susceptible to structural solutions, although none that have been proposed are without problems: regulating the media, for example, with regard to the balance or quality of their output, or subsidizing them not to keep them afloat, but to make them "independent" of interest and ideology.[37] What is particularly significant for present purposes is that some of these explanations for poor media performance imply that part of the problem stems from what *the public* demands.

Blaming the public for the deficiencies of the media is an old practice, and often not an unreasonable one. A magazine, a newspaper, a book, a television program, a blog can be effective only if enough people are willing to attend to it and think about it. Media that depend on subscriptions and advertising can stay in business only if they attract and keep an audience. Clearly, one of the principal reasons that the core public sphere is generally not effective in maintaining an adequate critical posture in wartime (or during a facsimile of wartime) is that the audience for its products will not tolerate much criticism of national policy. It is true that this lack of toleration is often promoted by the government, but the promotion would not work if there were not already a disposition on the part of the public to be intolerant under such circumstances. Of course, wars are not constantly being fought, but one of the most notable tendencies in public practice during the last several decades has been to try to replicate the conditions of wartime as far as public discussion is concerned, in part in order to stifle critical examination. We have a lot of wars going on: a war on terrorism, a war on drugs, a war on obesity. We once had a war on poverty. If we really wanted to do something collectively to improve the functioning of the public sphere by raising the level of performance of its audience, we could stop declaring war in any but the strictest sense of the term (and then only when necessary).[38] The institutions of the core public sphere could contribute to improving public discourse by criticizing the use of the "war" metaphor in other media, by politicians and interest groups, and by the government, and also stop doing it themselves. Universities could encourage their students to learn to recognize the practice for what it is and to appreciate its consequences.

When the deficiencies of the discursive public are discussed, they are typically described very broadly, essentially referring to everyone who can read. In the past, it was frequently argued by opponents of democracy that the "masses" are simply incapable of the level of rational thinking that is required for following sophisticated discussions or understanding the need for searching criticism. (As we will see, similar arguments were also made by people who claimed to support democracy.)[39] In contemporary society, however, we focus on education. If the people don't behave as rationally as they should and with sufficient information, it is assumed to be the fault of the system of public education. This is not wholly unreasonable. But one of the things that we seldom take into account is the fact that only a proportion of the citizens *choose* regularly to take part in the conversations of the core public sphere and to acquire the knowledge and skills needed for doing so. The group that chooses to participate in an informed way is by no means identical to the pool of university students and graduates. It is perfectly possible to take part in sophisticated discussion without certification of higher education, and many people do. A great many other people who have been to a college or university do not participate. (And anyway, as we have seen, the boundary between the conversations of the core public sphere and the broader discursive public—between, say, the world of *Harper's* and the *Atlantic* and the world of *People* magazine—is not clearly drawn. Some people participate in the second but not the first, some in the first but not the second, some in both, and a great many read *Newsweek*.) Nevertheless, it appears that the overlap between the group of people who have received a college education and those who choose to participate in the core public sphere (as readers, subscribers, and so forth) is fairly high.

Recognizing the element of choice is important from various perspectives, as we will see later. With regard to the recent performance of the core public sphere, it may indicate not so much a weakness in everybody's education as a need to focus on those who go to college. If we want to increase the amount of participation in the core public sphere, we should maintain and, if possible, increase the proportion of the population who go to college—remembering that one of the most significant practical dimensions of the modern democracy is large-scale engagement in public discourse. If we perceive a deficiency in the capacities of the audience of the core media, we could concentrate on the educations of the people who are in college or will probably go there, particularly with regard to their knowledge of the background of public affairs and their ability to analyze critically texts in the media. This might require a shifting of emphasis in the undergraduate curriculum. We could create strategies for encouraging college students and graduates to choose to participate in the conversations of the public sphere. We will discuss these and other possible actions in later chapters, especially Chapter 7. Note that, in terms of justifying support for public universities, they have nothing to do with economic improvement, employment, or increasing the quality of a student's life or value as an individual. They do relate to citizenship, but in a specific way, one that is directly connected to the general and liberal education elements of the undergraduate curriculum.

Let us return to the question of why subscription rates for quality newspapers and magazines have declined. Apart from the reasons that have been mentioned previously (the greater attractiveness of the internet, for one thing, and the currently costless availability on the internet of much of the news, information, and analysis that one would get by subscribing), there may be others that make sense in terms of the particular composition of the audience element of the core public sphere. It is clearly not because the pool of potential readers has been declining, since the number of college students and graduates has been increasing for some time. (The increase has been slowing and may level off or disappear, depending on what happens to financial support for higher education, but these factors have no bearing on declining subscription rates in the past.) College graduates have been *choosing* not to subscribe. This could be because of deficiencies in their educations or because they have not been encouraged to take part in the public conversation. It could also be because of a much-discussed decline in the interest of educated people in things having to do with the "public" in any of its forms: the effects of the "me generation" and its successors. But it may also be a result of deficiencies in the media of the core public sphere that are perceived by reasonable, educated people. It could be the case, for instance, that many of the quality opinion magazines are ideologically and discursively inward looking, so that instead of speaking to the concerns and interests of an educated general audience, they have become dependent on circles of people who think, write, and talk in the same way that their staffs do. (It seems to me that this is true of the *Nation* and the *National Review*.)[40] It is not that the circles are closed, but you have to know the languages being spoken and the assumptions that underlie the discussions in order to enter. These things can be learned, but you have to make the effort, which generally requires a prior commitment. Why, if you have no such commitment and no desire to learn the language, would you want to subscribe to a magazine that required them? Several other possibilities of this sort could be cited.

There is a great deal that could be said about the problems of the core public sphere. But it is time to turn to the second element in the relationship that is the central theme of this book: the problems of public higher education.

CHAPTER 2

Why Is Public Higher Education in Trouble?

People associated with state-supported universities have complained for many years about a "crisis of public higher education," either on the way or already upon us. It was once easy to dismiss much of this kind of talk. Some of it came from university administrators at times when state legislatures were debating appropriations. *Everyone* uses such language at those times. Some of it came from faculty concerned about what they perceived as a decline in grading standards and student performance or the inadequacy of intellectual fashions adopted by other faculty: natural conservatism, sour grapes, and the effects of the "academic ideologies" mentioned in the introduction. Faculty and administrators both expressed nostalgia for the "golden age" of public university expansion and relatively abundant funding in the 1960s, by comparison with which everything that has happened since can appear to be unrelieved decay: unrealistic perspective. From the political left came complaints about a decline of commitment to fundamental social change among students and about universities bending the knee to money and power; from the right, complaints about institutions encouraging an excess of such commitment and proscribing alternative viewpoints (specifically, conservative ones).

In the last few years, however, a realization that public higher education is in deep trouble has become very general, as has recognition that the trouble transcends the particular sources of complaint familiar from the past. The term "crisis" is now often used in its proper sense, to refer to a condition in which the subject faces a real possibility of extinction and that recovery, even if it occurs, will probably bring with it substantial changes. It is true that a crisis in its classic meaning is not necessarily a bad thing; the changes that accompany recovery or that make recovery possible may be desirable and necessary. For the most part, though, the current discussion of the crisis of public higher education portrays the impending changes as disastrous. It presents the crisis as a threat—not, indeed, to the existence of public higher education *per se,* but rather to its performance of important functions that cannot be handled by other institutions. The threat is very real. This book will attempt to point out opportunities that the crisis provides, but it will not make light of the alternative possibility: some of the essential features of modern American society, features that effectively make the United States a liberal democracy, will be severely crippled if the crisis of the public university is not resolved satisfactorily.[1]

The proximate cause of the current crisis of public higher education is financial. Since the economic collapse of 2008, the fiscal problems of state-supported universities have become quite severe. In most states, shortfalls in tax revenues have meant that public universities are facing significant, possibly crippling, budget cuts, just at a time when, because of the economic downturn, enrollment demand is high. (People who can't get jobs often use the opportunity to continue their educations.) The huge fall in stock prices has devastated the endowments of both public and private universities, although except for some of the leading public "research" universities, it is the private institutions that have been most directly affected by the markets. In most states, the immediate source of fiscal problems for state universities is falling appropriations.[2]

But why does this situation constitute a "crisis" in the strict sense of the term? There is no reason to think that appropriations for higher education in most states will not rise at least to some extent when state tax bases expand to their earlier sizes. The business cycle will inevitably swing back again, helping endowments. As I write, it already is doing so. There is evidence that, prior to the 2008 recession, the overall trend in state appropriations for public universities and colleges was on the rise, after a period of stagnation or effective decline between 2002 and 2005.[3] Why shouldn't we expect the trend to recur?

We have to place the financial problems of public universities in more specific and longer-range contexts in order to see where the elements of crisis lie. If we look, for example, at the evidence that appropriations rose just before the bust in 2008, it is notable that the increase was very uneven among states. A high proportion of the states registering large increases were ones with relatively poor histories of supporting higher (or any other kind of) education and without strong research-oriented private universities. These states were playing catch-up. For the most part, they were doing so largely because their leaders had been convinced that research universities promote economic growth. With a few exceptions, the states that have been and still are the leaders in public higher education, both in creating research universities and in developing systems that educate large numbers of students, fall well below the mean percentage increase. They were not, on the whole, experiencing any effective recovery before the 2008 crash. Most were not even keeping up with inflation. In some cases, notably California, the pattern for well over a decade has been one of declining real income from state sources that has, among other things, reduced the accessibility of higher education to students from low-income backgrounds. Almost everywhere (even in the states that have increased appropriations), there has been a continual rise in tuition and fees over the past 20 years, well in excess of the overall inflation rate. Although politicians, responding to complaints from parents and students, have voiced objections to this and sometimes enacted limits on tuition increases, they have usually been unwilling to deal with the underlying problem: the fact that appropriations have not kept up with expenses and that the proportion of the operating costs of state universities that comes from appropriations has fallen steadily.[4]

There are signs that, under the Obama administration, the federal government may finally be intervening to do something about these financial problems. An initiative has been enacted to deal (marginally) with one of the most fundamental problems of American higher education: the fact that federal aid to postsecondary students has, since the 1960s, been delivered primarily in the form of guarantees on student loans. Under new legislation, the government will make many of such loans directly, thus possibly cutting costs.[5] It will also spend a substantial sum to bolster the community college segment of public higher education, largely because of the economic benefits that are expected to accrue from the expansion of two-year offerings but also because community colleges are the institutions of choice for a large proportion of students from minority backgrounds entering higher education.[6] These are significant steps, but there is controversy about whether they are entirely in the right direction; among people who think that public higher education is in crisis (including the sponsors of the initiatives), there is consensus that by themselves they are not enough.

As a result of their financial problems, public universities are faced with a crisis of structure and function that has many facets. One example: unless something changes, many institutions are probably not going to be able to continue to admit large numbers of low-income students and at the same time adhere to the goal of remaining or becoming research universities.[7] Reducing the number of admissions overall is not an option if they keep the latter focus. To follow the research university model, most of them *must* admit large numbers of fee-paying and appropriations-generating students.[8] If appropriations decline, they will have to compete with each other (and with private institutions) for the students who can pay fees—at the same time that they are engaged in a similar competition to attract the best undergraduate students by offering them scholarships.

The fiscal aspect of the crisis of public higher education is not a sudden development, but one that has been building for some time and that intimately involves the whole conception of what state-supported universities are supposed to do. Public universities are undoubtedly going to have to make changes. If they do not make changes deliberately, changes will happen anyway, although probably not in a manner that would benefit either them or the public that supports them.

But why must the current financial problems of public higher education have such extensive consequences? Why can't the universities make a strong case for holding the line with appropriations for the time being and increasing them at least in proportion to costs later, when economic conditions are better? University administrators are, of course, trying to do this, although with unimpressive results. Some of the arguments that they make are unconvincing to the public (or at least to the public's elected representatives). One variety of argumentation—the one that emphasizes the university's economic role—can actually be self-defeating for many institutions and carries the further danger of narrowing the scope of public higher education to such an extent that vital functions may go unperformed. If we look a bit more closely at this and other arguments, we will see that what is lacking is a conception, both within academia and outside, that coherently reflects the full range of ways in which state colleges and universities actually contribute to the public good. What this means is that public universities cannot effectively deal with the financial crisis they face (i.e., they cannot make a continuously successful argument for adequate state support). It also means that, quite apart from limitations resulting from falling financial resources, they cannot fulfill adequately all of their actual responsibilities to society and to the public. You have to know clearly what you are doing in order to do it well. One of these responsibilities—mostly hidden among the current justifications of public higher education—is to support public discourse and especially the core public sphere. But before we discuss this responsibility, let us look at the more familiar ways of justifying public higher education.

Public Higher Education as Economic Engine

The justification most frequently and most loudly articulated at present focuses on the value of state-supported universities to the economy. This argument has become almost hegemonic in recent years. Of all the arguments, it is the one that, when used not only to justify appropriations but also to inform the ways in which the appropriations are spent, poses the greatest threat of the role-narrowing mentioned earlier. There is growing evidence that the threat is being realized.

Let me give an illustration from my own experience. A few years ago, I attended the inauguration of a new chancellor at an urban public university. Literally every dignitary who spoke at the inauguration (legislative leaders, the mayor, the president and the chairman of the board of trustees of the state university system, the local congressman, and the new chancellor) referred prominently—in some cases exclusively—to the economic role of the university. The system president said that the university was dedicated to "serving the people and the economy" of the state, while the chair of the trustees said that its job was to "enhance the economic and social well-being" of the state. The state senate president indicated that the university's task (the only one he mentioned) was "educating our workforce" and described the higher education appropriations process as "investing in our system." The congressman seemed to take a different tack when he described the university as providing essential services (suggesting the model of a public utility), but the only specific essential service he mentioned was enhancing the state's "competitive advantage in the global economy." Even the chancellor (not himself an academic, as is increasingly the fashion), who devoted the major part of his speech to the importance of the liberal arts, said that the task of the faculty was to educate the "populace and the workforce."[9]

I hasten to point out that I agree entirely that *one* of the tasks of a public university is to support the economy. Moreover, I am certain that none of the people who spoke at the inauguration would, if questioned on the matter, say that the economy is the *only* concern of state-supported higher education. I suspect, however, that in talking about other considerations, most would fall back (as the chancellor did) on a discourse composed of clichéd assertions about the value of education to individual character and self-fulfillment that are seldom critically evaluated and so worn out as to be largely ineffective in constituting an alternative to the discourse of economic function. Some might also mention the value of public higher education as a means of affording social mobility to members of minorities (mobility largely determined by economic value added to the individual worker), and possibly—although, perhaps surprisingly, none actually did—the function of higher education in promoting citizenship. But clearly, the predominant theme was the utility of the public university to economic development.

There are several reasons that public university leaders leap so readily to economic justification for their institutions. One is that they have come to believe that this kind of rhetoric is what sells their point. It is a self-reinforcing belief. The more they use the rhetoric, the more the people with whom they are dealing think that it is the correct way of envisioning the functions of universities. Who, after all, should know more about universities than their presidents, chancellors, et cetera? (Considering the backgrounds of an increasing proportion of chancellors and presidents, this could be viewed as a joke, although it isn't quite—yet.) And the more that people outside the universities who are involved with paying and making policy for them talk in the same terms, the more the representatives of the universities feel the need to do so as well.

A second reason derives from a phenomenon that some critics describe as the absorption of American higher education into the corporate business universe, or "corporatization."[10] This is to some extent an outgrowth of a tendency on the part of state governments and institutional trustees to require universities to seek a larger proportion of their funding from cooperation with businesses—especially through research contracts and arrangements for exploiting faculty research and development. There is, in fact, a good deal of evidence that for most universities this pursuit does not add much to the institution's net resources, but the belief in its efficacy is quite general.[11] In some ways, it is seen as a response to the decline of direct state support. Whatever the intentions behind the trend, its consequences have begun to be felt deeply within the practices and cultures of many universities. If the principal concern of a university's leadership and the central criterion for judging its overall success is the effectiveness with which it partners with businesses, then priorities within the institution necessarily adjust themselves accordingly. At many universities at the moment (mine, for instance), finding and developing corporate partnerships has taken a large proportion of the scarce money available for pursuing new initiatives.

Apart from the emphasis on commercially significant research and on corporate partnership, the teaching function of public universities has also come to be described primarily in terms of producing skilled employees for business—as was clear in the summaries of the inauguration speeches given earlier. This tendency has even gone to the point of occluding the more traditional quasi-economic aim of educating professionals, and it has affected the self-presentations of teaching personnel. If, for example, faculty or advisors in the liberal arts fields want to show why what they do is important, they often trot out statements by corporation human resources officers to the effect that companies need recruits with bachelors' degrees who can read, understand what they read, and write, and that these skills appear to be most developed in people with liberal arts degrees.[12] (Of course, the same human resources people will admit when pressed that they offer much higher starting salaries to engineering and business graduates.) When state university presidents want to show legislators or the public why appropriations for higher education should not be cut, they cite the number of jobs the universities directly or indirectly create and the needs of local industry for graduates.[13] These discursive habits have been accompanied by changes in behavior and nomenclature within university administrations. Heads of some campuses are now officially styled "CEOs" and often try to act the part.[14] University administrators strive to keep up with the latest fashions in business management, planning, and financial strategy (in the last case, very much to their cost in 2008–9).[15]

The movement toward what we will from now on call the "economic-technological" mode of discourse is clearly part of a larger, essentially global tendency to emphasis markets, market behavior, and classical market theory as the principal features of social life and the main sources of criteria for adjudging the success or failure of nonmarket institutions.[16] One result of this tendency has been the emergence of a challenge to the very notion of a "public" or a "public interest" in any general sense. Perhaps the economic disasters of 2008–9 have turned back part of the challenge, but it is certainly not dead. If there were no "public interest," or no public interest apart from the interests of individual employers, business enterprises, employees, and consumers, there would be no very compelling argument for public universities.

Apart from this general consequence of "corporatization," there are specific ways in which the tendency works against public higher education. For example, consider the practice of regarding education as a service, students as either products or customers (depending on circumstances), and faculty as service employees or producers of commodities. Although we often treat such analogies between a university and a business firm as impositions from outside higher education, they can also arise from within. It is easy to see how this can happen if one has ever attempted to change anything at a university or to dislodge one's faculty colleagues from their disciplinary and ideological fortresses. The appeal of a form of discourse, a system of organization, and a set of criteria all oriented (in theory, at least) toward producing results (in this example, providing a service) can be very strong.[17] And within limits, models from the corporate sector can be applied quite successfully, although the limits are fairly narrow.

Serious problems arise, however, when these limits are crossed: when the culture of business is taken out of its appropriate setting and absorbed too completely, believed in too uncritically. Students are really neither products nor customers, although there are some ways in which thinking about them in those terms can be useful. In relation to the university, they are *students,* which is an established category *sui generis.* They have other characteristics that do not correspond to "products" or "customers." They are also future (or present) citizens, employees, and members of the public in its many forms. Education can in some ways be considered a service, but the customers and the beneficiaries of the service are multiple and cannot be envisioned as members of a single category in the way that "the customer" can be in business. There is service to the student, but also to the state, society, the public, and employers. Problems also arise when we treat what universities produce as a commodity (usually described as "knowledge"). Knowledge, in the minds of students or in the form of research results, is not the same as a product on the market, although to a limited extent it can be turned into one—not a large enough extent, however, to define the category.[18] In these as in other ways, the movement toward a corporate, economic model for all higher education imposes a single, overly simple, framework on a reality that is immensely complex and is not fundamentally the same as the one in which the model was conceived. (It could also be argued that "corporate models" do not catch the complex realities of business either, but that is another matter.) Not only do bad decisions result, but the whole nature of education is threatened with a transformation that emphasizes only a limited range of the vital tasks that education performs.

Even as a rhetorical device, emphasizing the economic role of public universities in seeking and justifying support does not really work. It does give policymakers—usually those already well disposed toward higher education—an excuse for approving appropriations, an excuse that seems "practical" and that apparently avoids ideological considerations. However, even in most of the states in which the universities have gone farthest toward research collaboration with industry, the proportion of appropriated revenue to other forms has declined and is continuing to decline.[19] Actually, according to the logic embedded in the discourse, this makes perfect sense. If university research is advertised as something that benefits businesses and universities together, it is reasonable for state legislatures to expect that over time, the products of the collaboration should not only pay for the research itself but also contribute to lowering the taxpayers' share of the costs of the university. Also, if universities are the path to good jobs in the knowledge-based economy, why should the cost of maintaining the path not be borne by those who take it and benefit from it—those who get the jobs? Such arguments are frequently made—although not usually when students and parents complain about rising tuition and fees, even though that is the obvious result when the argument affects practice. Instead, blame is placed on "waste" or "runaway costs" in the universities themselves.[20] Because waste can always be found or inferred in any institution and because it is often a nebulous concept, it is relatively easy to use it to divert attention from other problems. (What? Professor Lavransdatter teaches only two courses a semester and uses the rest of the time *supposedly* doing research on medieval Norse literature? Waste! Make Lavransdatter teach more courses, or get rid of her and her useless courses, and costs will fall.) Where the individual-benefit argument usually appears overtly is in explaining why student loans should be preferred to direct student support (scholarships, grants-in-aid, and so forth).

The example also demonstrates another of the self-defeating features of the economic-technological argument: it privileges the forms of university activity that are conventionally regarded as having economic significance. (What is medieval Norse literature *good* for, anyway?) For one thing, the argument can lead to giving funding preference to programs and courses that supposedly produce skills directly applicable to employment (especially in sectors of the economy that are currently fashionable as the "waves of the future"—which is one of the reasons that universities produced far too many graduates in computer science and similar fields in the 1990s).[21] For another, it can lead to a similar favoring of research that is considered to be significant for technological development. It is true that universities and academics have become very adept at claiming economic relevance for a wide variety of fields, and also at playing shell games with categories at budget time—by, for instance, placing all research (including Northern European literary studies) in one budget category labeled "research," and then arguing for funding as though all the items in the category had to do with technology. Most academic administrators—even now—actually subscribe to views of the university that do not entirely conform to the models they espouse and are willing to play little tricks in order to reconcile the resulting inconsistencies and keep their academic constituents happy. Nevertheless, these techniques can only be taken so far. The privileging of certain fields in terms of their perceived economic value (even if the evidence behind the perception is questionable) can be seen in almost all universities these days, and particularly in state-supported ones.

The Academic Ideology

Very few academics (outside of a few technical or business fields) fully accept economic-technological discourse or the corporate model, and a great many of them think that the discourse and the model are completely wrongheaded. When pressed to explain what they and their institutions do and why society should support them, they tend instead to reach into the collection of beliefs that was called "academic ideology" in the introduction.

The term "ideology" is appropriate in a number of ways. The American version of the collection was, as we will see, put together in the process of constructing the academic profession in the United States during the late nineteenth and early twentieth centuries. The collection served, and continues to serve, as the principal way of modeling the practices and the ethic of the American university professoriate and explaining how the academic profession embodies values of the greatest importance, values that transcend those that involve mere material features of human life. The fact that many of its elements were murky and sometimes contradicted one another made it no different from any other ideology. The fact that, although the ideology represented itself as a "tradition," it was actually not very old, also made little difference. A large number of the traditions incorporated in ideologies are manufactured for the purpose.[22] For a long time—until the 1970s—the academic ideology was employed with considerable success in convincing the people who supply money to universities that its tenets manifested the truth about higher education. It succeeded even more spectacularly in informing the self-presentation and the career expectations of university faculty, as it continues to do despite the many changes that have taken place in American higher education and academia. It helped to create, justify, and maintain a professional career that was satisfying, respected, and, if not outstandingly well paid, then at least secure and not as all-consuming of time and energy as, say, law and medicine. Small wonder that academics have held to it tenaciously.

On the other hand, in some ways "ideology" is not quite right, either. The adherents of an ideology normally expect to have to defend it against other ideologies whose advocates directly criticize its tenets. Many of those who use or refer to the academic ideology tend to treat the assertions of which it is comprised as something like holy writ, exempted from critical examination by the profundity of the values that it manifests. Descriptions of the values themselves vary somewhat with the age, the political position, and the philosophical predilections of the describer. Some see them as comprising "civilization" (meaning, usually, Western civilization). Others regard them as essential to advancing society's pursuit of the truth, usually by means of research. Still others view the elements of the academic ideology as key features of the process of liberation (variously defined) in the modern world. Most aspects of the ideology, however, are generic. On the whole, people who accept the ideology do not expect to be asked to think critically about it or to defend it, and generally they are not. When economic-technological arguments are made for higher education by universities these days, the academic ideology is kept in the background, assumed to be valid but only brought up—if at all—as an afterthought. (See the description of the inauguration mentioned earlier in this chapter.) When actions are proposed or undertaken that appear likely to damage the interests of faculty or the operation of a university, it was customary in the past for the representatives of the university to explain the dangers in terms of the values contained in the academic ideology. That has largely ceased with regard to the public universities, where economic arguments are the weapons of choice (with all the dangers and baggage that were discussed earlier). But when academics who are not administrators confront fiscal adversity, they generally blame it on administrators, legislators, and others who, they claim, have no understanding of the "true" value of higher education as embodied in the academic ideology. They tend to attack economic reasoning (even the most obvious kinds, such as "There isn't enough money to pay for...") as a sellout. They generally do not, nor are they asked to, defend their position. They are simply ignored. Few of them, in my experience, are willing to take the trouble to go beyond the accusation stage.[23] The implication is essentially that the position on their side is self-evident, and if the people on the other side can't see it, they aren't worth responding to. This is not a very effective strategy.

What actually *is* the academic ideology? Most of it is quite familiar: a collection of statements and implications put together from various sources and often not tied together very well. One component: the highest function of an academic is to increase humanity's store of knowledge and understanding of the world, and therefore the highest function of an academic institution is to foster that endeavor. If the institution in question designates itself as a "research university" in the classical sense of the academic ideology (which is *not* exactly what is meant by the term in the current economic-technological discourse of higher education), then the faculty's primary task is cutting-edge research. In the postmodern era, this notion has been extended without much difficulty to include interrogating the basis on which claims to truth are made and questioning the possibility of any such claims. (Logically, one might find an inconsistency here, but it has not had much impact on such university processes as considerations for tenure.) Most American public universities present themselves at least to some extent as research institutions, whether or not they are officially classified as such in the Carnegie Foundation system or according to state mandates. Some of them, it is true, either by their own choice or because of state enactment, require their faculty to concentrate more heavily on teaching than on research. Nevertheless, even in those cases, the most prestigious, most advertised sign of good undergraduate teaching tends to be former students who are accepted into graduate programs at leading (i.e., research) universities and who become researchers. Research is king.

The academic ideology specifies two principal conditions that designate appropriate research: it should be "objective" and it should be undertaken "for its own sake." (Again, a postmodernist qualification has to be made, since objectivity in its classical sense is not generally a possibility in that discourse.) These conditions are frequently described as being inconsistent with the notions of research embodied in the economic-technological model. They also incorporate conceptual problems of their own. What does it mean, for example, to do something "for its own sake?" Does potential utility rule something out as an object of academic or research interest? Or does the term refer not to the subject but to the motivations of the individual researcher? Is it acceptable to do research on a potentially lucrative topic if your main reason—at some point—for initiating the research was curiosity? How is anyone to tell—including yourself? And so forth. Objectivity is so problematical a concept that it would take a book just to address it here. It is not that the issues surrounding objectivity in science and scholarship are unimportant; it is just that they are complicated, as they are also in the practice of professionalism. They are not straightforward as they are represented as being in the academic ideology.

Some of the other elements of the ideology we can just list. Higher education aims not just at adding to knowledge but also to civilization. (This aim famously requires one to indicate what is meant by "civilization," a substantial bone of contention in the era of culture wars.) The function of adding knowledge and the important role of criticism in the process require that academics be protected as completely as possible against repercussions for anything of an academic nature that they do or say. They must have as complete freedom as possible to choose what projects they will work on (and, within the limits of their professional qualifications, what they will teach). These components of what is called "academic freedom" were imported from Germany over a century ago, together with tenure (although, as we will see, the principal justification for tenure was different from the one usually given today in the context of academic ideology). They entail a great many problems, both in theory and in practice. For example, what should a research-oriented university do when a faculty member ceases to do much research or to publish after attaining tenure? This is not unusual and has led to (admittedly exaggerated) criticism of universities in recent years.[24]

With regard to the pedagogical functions of higher education in the academic ideology, the university is held to "mold" the minds and characters of individual students, making them in some way better than they would otherwise have been. It is, admittedly, difficult to describe other than metaphorically how education "molds" anyone, as distinct from helping to supply ascertainable bodies of knowledge and observable skills. Fashions in this area have changed since the nineteenth century, but the desired traits that have remained more or less constantly cited are honesty (somewhat narrowly defined these days as avoidance of plagiarism), ability to work hard, and commitment to appropriate human ideals, as well as capacity to think critically and to argue civilly. (The last two, it seems to me, are really skills, not traits that can be shaped or molded.) The university and its faculty should uphold the strictest standards of student performance, for purposes of both admitting new students and assigning grades to the students they already have. Strict standards are the primary means by which the character of the student is shaped. In the academic ideology, moreover, a university education opens up to students the highest and most interesting vistas on human endeavor and provides the means by which good and ill fortune can be looked upon with equanimity. This list goes on; much of it is very familiar.

One of the most notable features of the ideology is its insistence on what might be called "academic self-referentiality." This is the claim that, in matters that are clearly academic (which are to be identified by academics), academics alone should be the judges of what is significant and what is not, what is done well and what is done badly, what constitutes appropriate scholarship and teaching, and what priorities should be—including priorities for distributing resources within an academic context. In collectively making judgments on these matters, academics are supposed to be responsible only to themselves, not to the state, to the public, or even (in the case of teaching) to their students. The principal justification for the claim about responsibility is that academia as a collective enterprise stands for things that transcend mere state or public: civilization, humanity's highest aspirations, et cetera. At the same time, the extent to which standards should be enforced even by academic collectivities on individual scholars (except for students) should be highly circumscribed, for fear of trespassing on the academic freedom on which the pursuit of knowledge and truth depends.

The academic claim to self-referentiality parallels ones made by other professional groups, especially those that possess the legal authority to certify their own members and to police their members' practices: for example, law, medicine, and accounting. In most of the parallel cases, however, the claim is largely based on assumptions about professional ethics and about the necessity of maintaining current professional knowledge, as well as on the supposition that self-policing will be accomplished diligently because the status and incomes of members of the profession depend on the public's perception that standards in the profession are high. And even so, self-regulation is not absolute: the state intervenes, sometimes very heavily. The academic claim is much more extreme, because the existence of neither a public *interest* nor a public *responsibility* is given much overt recognition in the academic ideology. It is simply presumed that in general, the world or the nation is a better place than it would be otherwise if scholars are able to do what they do—according to their own notions of what they should be doing. To all intents and purposes, the state does not intervene in the certification process for academics. Nor does it police professional (as opposed to personal) ethics—except occasionally under political circumstances that almost always evoke complaints about violations of academic freedom. One of the classic exponents of the academic ideology in the

United States, Abraham Flexner, was adamant about the need to separate academia as far as responsibility is concerned not only from the state, but from society and its interests in general.[25] According to Flexner, scholars and social scientists, driven mainly by their own curiosity, should study the problems of society from their detached location in the university and make the results of their inquiries available, but they should not take any responsibility for formulating or pushing for particular policies—apparently because fear of being held responsible would deter them from thinking things completely through.

These and other elements of the academic ideology taken together can be used to compose a comprehensive justification for supporting universities, including public ones, but it is one that requires that the people to whom it is addressed already accept most of its basic premises. The justification goes something like this: the essential feature of universities is their faculty. The faculty carries on a large part of the task of maintaining civilization, seeking new knowledge, and imparting both knowledge and civilization to younger people: high tasks for which no detailed justification is needed. University faculty must therefore be supported. They must not be required more than is appropriate to contribute to their own support by, for instance, directing their research toward money making. If they did that, the purity of the motives that lead to research would be violated and the objectivity of the researchers would be threatened. Similarly, the faculty must determine the curriculum of the university, among other things in order to ensure that students are educated according to the same principles and in the same fields that inform research. Curriculum should not be constructed according to the vocational desires of the students or the needs of employers for workers with particular skills. (Even in the case of professional fields, the presumption is usually that the faculty undertakes basic research and teaches not so much specific knowledge and techniques as the theoretical framework that underlies them.) The state or private donors should provide adequate funding for university teaching and research but must not expect a financial or practical *quid pro quo*. Moreover, because the best research is done when researchers enjoy the greatest freedom to choose what to work on and the greatest freedom from responsibility, it is entirely (or at least largely) up to university faculty to decide what is done with the support that is provided. It is also up to the faculty acting collectively to decide for itself whether or not the university and the individuals in it are performing well or badly. Altogether, the academic ideology represents universities as being properly dominated by their faculties and largely self-referential. This representation is supposed to govern the relationship between higher education and the public entities that pay for it. A slightly different rendering is used to deal with those who pay student fees, but it comes down to the same thing: we will tell you what it is that students need to learn and you (or whoever supplies the money for student aid) will pay what it costs.

When it is expressed this way—which is admittedly oversimplified but not fundamentally inaccurate—the academic ideology may seem a very unlikely basis on which to build an argument for public support for higher education. The surprising thing is that it has ever worked at all for that purpose, and in fact, as we will see in a later chapter, it has never worked entirely by itself. To be effective, it requires concurrence with its principal postulates on the part, at the very least, of elite policymakers, typically people who have taken in those postulates in the course of their own university educations, who think of themselves as being in some way members of a larger intellectual community to which university faculty also belong, or who are perhaps seeking a kind of personal validation or redemption from the sin of financial success by supporting a way of thinking and a way of life that reject such success. Universities have worked very hard to develop such concurrence. It has, however, become more difficult to do this in recent years, especially for public universities with respect to their appropriated funding.

The problem is not, I think, that the major tenets of the academic ideology are in some objective sense false or valueless. It probably is true, for example, that society in general is well served if researchers have the means and the freedom to pursue any direction of inquiry they wish and to publish whatever they like. Indeed, I think that one of the defects of the way in which the ideology is embedded structurally in universities—in departments that are, at least in the arts and many of the sciences, constructed around disciplines that were defined over a century ago—is that it acts in practice to limit the variety of paths that can be taken in research and teaching. Nevertheless, I cannot *prove* the benefits of unfettered research in any definitive way (although I could cite examples that more or less support the concept). To assert the desirability of freedom of research is essentially to express a value implicit in the modern notion of the university, a statement that is consistent with a certain amount of experience and with some of the broader values of our society (which I share). This is fine; the values are admirable and the experience impressive. At the same time, there is no particularly convincing reason to infer from them the absolute precedence of "pure" over applied research or of research undertaken to satisfy the curiosity of the investigator over research intended to solve a practical problem that might produce a profit. (Of course, the reverse is also true, as we discussed earlier. If research for profit takes precedence over all other kinds, a great deal of interesting and useful work will not be done.) Nor, for that matter, is there a very strong case to be made that decisions about what research to fund, what research is useful, and what research is good should only be made by academics. Clearly, they should have an important voice in these matters and experts in particular fields must be listened to, but experts are just as capable as anyone else of bias, misinterpretation, and interest; they tend to have interests that are peculiar to themselves. (One point that could be made is that, ideally, these are matters for discussion in the public sphere.)

Thus, at least two of the characteristics of the research aspect of the academic ideology—the insistence on "purity" of motivation and on self-referentiality—are not as strongly grounded in logic or practical experience as the notion of freedom is and are more difficult to defend when objections arise. Even the idea of freedom to do research can easily be exaggerated, as both a proposition in itself and a doctrine that should be observed by suppliers of public funds. Some research is bound to be harebrained in conception; it is acceptable to permit a certain amount of it simply because the boundaries of hare-brainedness are difficult to draw and—occasionally—research that seems flakey turns out to be innovative and even world changing. But this cannot be an absolute or a boundless permission. Decisions do have to be made (and are made) to cut off funding for pointless projects because funding is never unlimited. Moreover, neither the public nor donors nor universities can be expected to pay for unlimited amounts of research on arcane subjects that are interesting to only a few people. Again, decisions have to be made, and they are. In its more absolute articulations, the academic ideology does not fit many of the realities of university practice all that well. The same points could be made about other aspects of the ideology besides its treatment of research.

The academic ideology worked very well as a basis for building an academic profession and for creating in the modern university an organization of great efficiency for generating intellectual enterprise of the most varied nature—as long as the varieties conformed to the boundaries of university departments. It has provided a cultural framework for insisting on the validity of areas of human endeavor that cannot be fitted into markets, which is a very important accomplishment indeed. But as a way of justifying support for public higher education on the scale that is now required, as a way of countering the hegemony of the economic-technological discourse in higher education, it simply does not work. This may be due, as many of my colleagues complain, to a decline of civilized values in the contemporary world, but I suspect that other factors are at least equally important. The very extent of modern higher education, the number of students, the number of instructors and researchers, the overhead costs of all aspects of higher education, all of these things mean that responsible policymakers simply cannot even try to underwrite the whole enterprise on the basis of the tenets of the academic ideology. This is a particular problem since, as we have seen, a large proportion of public colleges and universities represent themselves as research institutions, which leads their faculties—if they do not want to portray themselves in terms of economic function—to expound notions of research (derived from the academic ideology) that are convincing mainly to themselves.

There is also another factor: we can no longer expect either policymakers or the part of the general public that concerns itself with higher education to be willing to accept uncritically the assumptions that underlie the academic ideology. A substantial proportion of them identify academic claims to autonomy, academic freedom, and self-judgment as means by which a minority of intellectuals demand a tax-supported platform to propagate a particular political stance. This is by no means new, as we will see. In practice, the leaders of American universities in the past have had to balance this view (not infrequently subscribed to by trustees and legislators) against the tenets of the academic ideology, holding, for instance, that the latter were acceptable as long as they were applied in such a way as to exclude radicals.[26] At present, however, the phenomenal spread of conservatism among the public at large, closely linked to the spread of the idea of the market as the appropriate framework for thinking about everything, has made even that balance untenable.[27] The academic ideology is either rejected or ignored in a large sector of public discussion.

One of the weaknesses of some of the noneconomic arguments made by supporters of public higher education is that they tend to present the elements of the academic ideology as self-evident truths that require only a clear exposition to compel acceptance; they also tend to ignore inconsistencies between elements. What they should do instead, I suggest, is to explain clearly and in terms that do not depend on prior acceptance of the assumptions underlying the academic ideology just what the *public interest* is in higher education.[28] They need to explain why taxpayers should provide large amounts of money for the salaries of faculties that claim to be responsible only to themselves and to nonspecific entities such as "civilization," whose members describe themselves primarily as researchers free to pursue any topic they wish, no matter how obscure, and that resist supplying the kind of vocationally oriented education that large numbers of their students claim to want.

Utility to the Public

There are ways of doing this that do not simply reiterate the academic ideology, that put forward claims that state-supported higher education is useful in specific ways to a collective national public. Public universities can, for instance, justify themselves as promoters of good citizenship. The argument, however, applies mainly to the teaching function. There are, moreover, several definitions of what a good citizen is, most not very specific and some probably inappropriate for higher education. "Citizenship" is often regarded as something to be learned *before* one gets to a university. Universities can more convincingly claim that they produce public *leaders,* although most university degree curricula do not in fact have a specific place for leadership except in some professional programs. "Leadership" tends to be part of professional curricula, and there is some question as to whether it can actually be taught at all. Universities can also advertise the role of their research institutes as suppliers of information useful for governmental decision making. But research institutes do not have to be attached to universities in order to function and are usually supposed to be self-supporting, so arguments for public funding of universities cannot make much use of them as examples. These approaches are, I think, heading in the right direction in that they address the question of public interest, but they are far from adequate.

There are also arguments based on the proposition that universities serve a public function by providing social benefits, usually described in terms of social classes and minority groups. Public universities have made it possible for people from less affluent or disadvantaged backgrounds to qualify for professions and highly skilled employment, thereby complicating the distinction between "middle class" and "working class" in the United States.[29] They have served as one of the prime avenues through which excluded groups have entered the mainstream of national political life (although it will be argued in Chapter 5 that some of the most important ways in which they have done this have been neglected). These are real and vital accomplishments. Any attempt to develop a comprehensive justification for public higher education in the United States must incorporate its actual performance of what we can call the "social-democratic" function.

There are, however, several problems that arise in doing so. One is that the social-democratic function has generally been performed in a state of tension with the academic ideology. The latter includes an emphasis on standards that has only a very limited place for the broad-scale university admissions and the various curricular devices for responding to "substandard" university preparation that have made it possible to accomplish the former. For well over a century, there has been a constant current of complaint among academics in public universities about the terrible undermining of standards, about the selling out of the proper role of the university that result from excessive admissions of "unprepared" students.[30] Starting in the 1950s, inconsistencies between the academic ideology and the social-democratic task of the public university were to some extent papered over by the wholesale adoption of the meritocratic model that has become one of the most distinctive public faces of American higher education and also (as we will see later) one of its greatest curses.[31] (One aspect of the curse that is relevant here is the way in which the meritocratic model promotes the hideously dysfunctional practice of assigning quality categories to institutions on the basis of their alleged "selectivity.") The idea that universities order the students they seek to admit along a dimension of intelligence, accomplishment, and promise but without regard to class, ethnicity, race, and gender, and then offer them admission starting at the high end of the dimension, appears to do the trick: the practice asserts standards and identifies the "best" students, while at the same time supposedly not distinguishing between people on the basis of anything except relevant personal qualities.

It is, of course, exactly that: a trick, based on an illusion. There is no such thing as a single dimension of qualification and no way of constructing one that does not require manipulation in order to correspond to the complex social and political realities in which universities exist. Manipulation has occurred for various reasons: to admit students good in one subject but poor in most others, to make sure that the children of alumni and wealthy people are admitted, to improve the football team, to meet admissions goals for minority students. Unfortunately for public universities attempting to perform their social-democratic role, the last of these categories of manipulation (not the first three and a number of others) has been treated by conservative (and not-so-conservative) groups in recent years as a violation of the principles associated with meritocracy.[32] Manipulating to give precedence to minorities in admission at what are held to be the more prestigious universities is assumed to mean that "better-qualified" nonminority candidates are being passed over. This position is easily associated with the academic ideology's criticism of nonselective admissions and its insistence on standards, which enhances the conservative position's apparent legitimacy and allows those who propound it to accuse universities of practicing hypocrisy by not adhering to their own principles.

This brings us back to the problem of trying to justify support for public higher education on the basis of its social-democratic function. Considering the way the latter has come to be embedded in a discourse of meritocracy, even raising the subject threatens to embroil anyone who suggests it in more controversy than it often appears to be worth. No wonder that, on the whole, universities these days choose to present the argument only in its most attenuated form, one that can be easily slipped into the economic-technology argument: by opening professional career paths, public universities are creating opportunities for individuals regardless of social background, and thereby, by inference, reducing one source of social disparity. If one accepts that a hierarchy of prestige exists among institutions of higher education, it can be assumed that the less appreciated of the public colleges and universities will offer these services to the "less-qualified" students, including members of most disadvantaged minority groups.[33] But since few of the latter institutions are willing to acknowledge their low status (except as a burden that they need to throw off by becoming more selective), not many are in a position to emphasize their roles in this regard in justifying their existence. And the more-prestigious universities obviously do not do so, either. So the utility of even the most tepid version of the social-democratic argument for public higher education is, in practice, quite limited.

Christopher Newfield has recently made a more ambitious social case for public universities: the social-democratic approach on steroids. He sees the current crisis of public higher education as the result of a conspiracy of economic and social elites to head off the formation of a class of university-educated knowledge workers that would, if fully constructed and appropriately educated in public universities, amalgamate the American professional middle class and the working class and wield preeminent power in the United States. He postulates that this power would be used to produce a truly egalitarian society.[34] Newfield dates the beginning of the conspiracy to the late 1960s, when conservative elites realized that, mainly because of the enormous growth of public higher education, the new class and the new egalitarianism were becoming realities. He describes the resulting scheme as encompassing the corporatization of higher education, a huge reduction in funding for state universities, an end to affirmative action admissions, and regulation of the content of university teaching and of public statements by faculty. One of the key features of the conspiracy is an effort to frame wedge issues that turn the components of the potential amalgamated class of the professional and working educated against each other. This, Newfield says, is the significance of the attack on affirmative action in university admissions.[35]

That there has been an attack on public higher education by various elements of the ideological right in the United States would be difficult to dispute, especially in California, the state on which Newfield concentrates. That the attackers perceive the attack in terms of the specific class fears that Newfield describes is less clear, although I suspect that he is right about a significant proportion of them. But even granting a cabal of elite conspirators, some important questions remain. How is it that the conspirators have managed to obtain the support of so many other people, including those in groups whom Newfield indicates are the principal beneficiaries of the policies under attack? Why do supporters of universities find it so difficult to respond effectively to the attacks?

Part of the answer to the first question is that a large portion of the college students and graduates whom Newfield sees as the bearers of the new egalitarianism do not, as Newfield admits, actually subscribe to it. His response to that problem is that we need to convince them to do so as part of their educations, by exposing the reality of class conflict in America and providing students with the theoretical equipment that will allow them to understand the extent to which the political right has imposed on them. That is all very well, but the argument, by postulating that there is only one way of looking at these things and that universities must present it, essentially admits to the intention of using public higher education for ideological indoctrination, for enforcing "political correctness"—one of the principal accusations that conservative critics of higher education make and one of the bases of their argument for limiting expenditure on public universities. In the form in which Newfield presents his argument, it is not likely to be successful.

Newfield's reply to the second question—why don't public universities respond to criticism effectively—is primarily that they do not try hard enough and that they are caught up in the logic of the economic-technological approach to explaining their function. As we have seen, the latter point is quite accurate. So, probably, is the first, although hopefully if public universities follow the lead set by Newfield and a few others, this will not be the case for long.[36] But a vigorous case for public higher education has to be built around a convincing argument, and none of the ones described thus far is likely to fill the bill, at least by itself and probably not in combination with any of the others. Newfield tries to supply one, a more specific version of his general call for public universities to teach their students to recognize the realities behind conservative scheming. As the foundation of his case, he cites the utility of research in the humanities for producing cultural knowledge that liberates its possessors and promotes social progress.[37] The arguments do not seem to be entirely convincing to me, and I doubt that they would convince the public at large.

For one thing, Newfield is particularly concerned to show the public value of the kind of postmodern cultural criticism that has become the leading edge of research in literary studies in the past few decades. He suggests that what the conservative opponents of public universities have *really* been worried about is the capacity of postmodern criticism to undermine the assumptions on which the existing social order is built and to provide a common framework for the projected egalitarianism of the university-educated middle class. He cites especially the work of Jean-François Lyotard as an example of the kind of theory that could be used for this purpose. I find this very difficult to agree with. I doubt that more than a handful of the opponents Newfield is concerned with have ever *heard* of Lyotard or are, in general, really haunted by the specter of postmodernism (except in the form of cultural relativism, which is not unique to postmodernism). They seem to be far more worried about more conventional positions on the political left. Nor do I see how a group of writers whose language is so esoteric and whose outlook on society and politics is so incoherent, who are in many cases critical of coherence itself, could provide an effective common framework for any kind of social change or movement whatsoever—nor do I see any evidence that they have. What conservative critics obviously *do* see in postmodernism is a wonderful source of red herrings to put before the public. Even if one were to grant Newfield's thesis of a conspiracy as the root of the problems of public higher education, building a case against the conspirators around the claim that public universities are performing a vital service by fostering cultural criticism is going to persuade, at most, only a relatively small number of people: those, say, who know anything about Lyotard. Significantly, most of those people are located on university campuses—although they are only a minority even there.

The last point brings up another serious problem with Newfield's approach to justification: his argument is almost entirely couched in terms of public *research universities,* of which for him the models are the campuses of the University of California. The argument projects the outlook of the faculty (for the most part, the humanities faculty) of such institutions, and in particular a belief shared by many of them that whatever they do—which, as the institutional category name implies, is mostly research—must necessarily be vitally important to society as a whole. This is essentially the point of the "academic ideology," updated for the early twenty-first century. Postmodern cultural and literary criticism is an example. It is centered overwhelmingly in university humanities faculties (and in certain other places in the core public sphere) and has at most a modest resonance outside, but is assumed within groups so located to be immensely significant. Not surprisingly, these groups are frequently disappointed when they perceive the unwillingness of the general public to support them unreservedly and its inability to understand what they are saying. Newfield largely ignores the majority of public colleges and universities that are *not* research universities in the sense claimed by the University of California components. One of the difficulties that American public higher education faces is the prevalence of the research university model as the ideal for all higher education. If there is to be an effective argument for supporting public universities, it will need to be one that encompasses most or all of them, not just those that fit into the research university category.

It might appear, then, that public higher education is in a pickle; and so it is, if its reasons for demanding stable or increased support must be based on the arguments that we have discussed in this chapter. It is not only a question of finding the correct rhetoric (although that is, as always in public affairs, a not-insignificant part of what must be done). The forms of justification that we have discussed all represent realities, albeit of different kinds. The economic-technological argument reflects the reality of corporatization and the harnessing of the university to the technological demands of the modern economy and of the national government as global power. This is a reality that probably should be altered, but it is not going to happen soon in any comprehensive way. The academic ideology reflects a reality deeply ingrained in the attitudes of faculty and in some of the practices of their universities, however poorly it might correspond to the other functions the universities actually perform. The social-democratic argument reflects many of those other functions, often in apparent contradiction to the academic ideology but in consonance with demands from major components in American society. Arguments like Newfield's, which in a sense combine elements of the last two, are based on the real (and impressive) existence of the research university, both as a model and as an ongoing enterprise. A change in rhetoric will probably not be sufficient; it will have to be matched, at least to some degree, by changes in the realities in which the various current rhetorics are embedded.

There is, I suggest, an approach to explaining and justifying the role of public universities that would work better and yet would not be wholly inconsistent with most the others. This approach was suggested when we discussed the idea of public universities as sources of training for citizenship and leadership in a *democracy*. By itself and in the usual forms in which the idea is presented, it is probably not adequate, as we saw. What is required in order to make it so is to recognize that the *public sphere* is a crucial element of democracy, that one of the realities in which universities operate is that they support and participate in the *core public sphere* described in the preceding chapter, and that one of the central tasks of public universities has been, and remains, the *democratization* of the core public sphere. The core public sphere is also in trouble. If public universities could promote an understanding of the importance of the core public sphere and of their own centrality in it, if they could alter some of their practices to emphasize their connections to public discourse and to support it more effectively, and if they could contribute to reconstructing the public sphere to meet new circumstances—all of which they are entirely capable of doing—they could make a much better case for themselves while helping to build a more competent American public. To see how this could happen, we have to look first at the history of the core public sphere and its relationship to higher education in the United States.

CHAPTER 3

Building the Core Public Sphere

The eighteenth-century public sphere described by Jürgen Habermas had little to do with universities. Except in a very few countries, the eighteenth century was not a golden age of higher education in Europe. The Enlightenment was primarily a phenomenon of written, printed, and oral exchanges among intellectuals that occurred outside the walls of educational institutions—in newspapers, journals, and books in vernacular languages; in coffeehouses; in salons; and in other informal locations. Except in Germany and Scotland, very few major (or even minor) Enlightenment figures had significant attachments to universities apart from attending them—and far from all had done that. The venue of the Enlightenment, at least in Britain and France, was Habermas's public sphere.

The eighteenth-century public sphere was a social space, mapped out in terms of structures and norms, practices and attitudes. It was, according to most of the historians who have studied it, a real thing, an existent social and behavioral entity. It was also, however, an ideal, although the eighteenth-century people who portrayed it as such had no specific name for it except "the public." It was wider by far than the entity they called the "Republic of Letters." Early in the century, the publishers, authors, and editors of the *Spectator,* Joseph Addison and Richard Steele, deliberately tried to create in the pages of their journal a model of an English public sphere as a site for discussion and as a set of desirable discursive practices.[1] The model, further elaborated by other writers and by usage throughout Western Europe and America in the eighteenth century, imposed itself as a central fact of life on both sides of the Atlantic. Habermas cites the coffeehouse as the structural archetype of the public sphere, but other private establishments, such as taverns and clubs, were equally important.² Periodical printed media also defined the public sphere, as both adjuncts to coffeehouse conversation and vehicles for expression of opinion in their own right. The periodical press made it possible to have virtual conversations, to create continuous webs of discussion within the realm of consensus developed in the public sphere. The press not only disseminated ideas and information, but also stimulated and supported intellectual intercourse among readers. So did books and collections of essays.³ The effect of the eighteenth-century public sphere on European, Atlantic, and indeed global history was immense.

Nineteenth-Century European Public Spheres

We are not, however, concerned here with the eighteenth-century public sphere, but rather with the constructions that succeeded it in the nineteenth century. "Constructions" is the appropriate term. They were very conscious creations, built around institutions that were much more resilient and permanent than the coffeehouse coteries of the eighteenth century. The most important of these institutions were periodical publications, voluntary associations, and political parties (at a time when European parties consisted of collections of voluntary associations not unlike clubs). The people who worked in and around such structures built a distinctive culture—primarily a print culture, but one that, regardless of the ideological orientations of its bearers, incorporated a relatively homogeneous set of frameworks for oral discourse, for analysis, and for the making of careers through literary production. The early nineteenth-century public spheres of Britain, France, Prussia, Austria, Italy, Spain, and Russia—in fact, practically all of Europe—constituted something altogether more formidable, comprehensive, and politically effective than the structures Habermas describes in the eighteenth century.

The central fact with which all European public spheres engaged themselves was the nation-state, modernized as a result of the Napoleonic Wars but not democratized.⁴ In countries such as Britain and France, which were politically united and possessed most of the institutions of modern states, the public sphere oriented itself principally around the issues and conflicts of national politics and the modes of national culture. In Germany and Italy, made up of (more or less) modern states that were not united politically, the *prospect* of becoming a full-fledged nation-state was one of the central subjects of public discourse; the public sphere functioned *as though* the united nation already existed—which was one of the reasons that, by the last quarter of the century, it did.⁵ In Russia, the public sphere was highly circumscribed in scope by the policies of an authoritarian regime and by the fact that the state was not, in a Western European sense, fully modern, but within those limits it was highly active and brilliant. In Russia, however, it was not closely linked to the structures of politics and government; its personnel were isolated into an *intelligentsia*.[6] In Western European countries, on the other hand, a key feature of the public sphere was its close connection to the structures of politics and administration as well as to the cultural and, often, economic elites. Such linkages had existed in the eighteenth century, but they became much stronger and more institutionalized in the first half of the nineteenth. This is one of the developments that leads Habermas to claim that the sphere disappeared in the nineteenth century, because he postulates the need for substantial separation between public sphere and state for the former to function properly.[7] It would, I think, be more accurate to see the nineteenth-century Western European public sphere as an adaptation to new circumstances, with both strengths and weaknesses in comparison with Habermas's ideal type of a public sphere that observes the state largely from outside. We can call this adaptation the "core public sphere."

A crucial feature of the European core public spheres of the early and mid-nineteenth century was the fact that it was possible to make a respectable career in them, and also in some cases to parlay an early career in the public sphere into a later one in politics or some other field of endeavor. Some of the leading figures in nineteenth-century British politics did so. Thomas Babington Macaulay first made his reputation as a contributor to the *Edinburgh Review,* the leading moderate liberal magazine of the era, and went from there to the House of Commons and to office in several Whig governments. Lord Salisbury, the Conservative prime minister at the end of the century, started as a writer and editor for the Tory *Quarterly Review.* Benjamin Disraeli began as a writer of novels that focused on social issues. (Novels were a central medium of public discourse on major issues in the nineteenth century.)[8] This was not the only way to get into politics, but it was a common one. Journalists who worked professionally for newspapers as opposed the quality reviews, monthlies, and quarterlies found the door to politics less widely open to them until the twentieth century, but once politicians and journalists discovered how much they needed one another, they tended to interact with increasing regularity. Only gradually did high-class journalism become a profession in the modern sense in Britain, but in the nineteenth century it was a considerably more structured way of life than that of the gentlemen-contributors to eighteenth-century public discourse and the Grub Street hacks who produced Georgian popular journalism.

The core public sphere in France was similar, except that the relationship between political parties and quality newspapers and magazines tended to be organized somewhat differently. Whereas in Britain, magazines and newspapers operated separately from the parties, despite often having obvious partisan leanings, for most of the nineteenth century in France the editorial staffs and major contributors of magazines and newspapers essentially *were* the leaders of parties.[9] In Britain, leadership of both major parties tended to be exercised by members of the well-established (and often aristocratic) social elite, who associated with journalists, editors, businesspeople, and civil servants at London clubs, of which there were no exact replicas in other European countries or in the United States.[10] These clubs were to some extent outgrowths of the kinds of structures in which the eighteenth-century public sphere had been situated.

The nineteenth-century Western European public spheres were clearly, by our standards, elitist; the media at their centers were published for and written by an educated minority. On the other hand, the educated minority of readers and participants in discussion was not limited to traditional hereditary elites, to the wealthy, or even to the ill-defined social class whose male segment called themselves "gentlemen." Although few of those who commented on the new quality media thought of them as "democratic," few referred to them as projections of the upper classes either. Any literate person who wanted to do so could read the quality journals as well as novels, newspapers, and works of history (a large proportion of which were written by people, like Macaulay, with political aspirations). By the end of the first quarter of the nineteenth century, the circumstance of being "respectable" meant that one regularly read material of significant intellectual quality, if only a good newspaper and the latest novels with aspirations to being "literature." Moreover, although the writers and editors of the central media of the public spheres tended to share similar educations and behavior patterns, their actual social backgrounds ranged fairly broadly: relatively few people from the poorest classes and the peasantry, but a significant number from nearly every other part of the social order. The public sphere was, as Habermas describes it, a fixture of "bourgeois society," which meant not that it was made up entirely of people from the "middle classes," but rather that it incorporated people from a wide range of family backgrounds, income levels, and political positions who shared similar educations and cultural modes that can collectively be called "bourgeois." People as different in social origin as Macaulay, Disraeli, and Lord Salisbury began their careers in the public sphere. So did Karl Marx.

The most prominent structural feature of the nineteenth-century European core public spheres was what we can call the "quality" press, which consisted of a few newspapers and several magazines (reviews, weeklies, monthlies, and quarterlies) in each major European capital. In Britain, the liberal *Edinburgh Review,* the radical *Westminster Review,* and the conservative *Saturday Review* and *Quarterly Review* became the models for a host of other periodicals throughout the English-speaking world and the effective centers of Atlantic public discourse.[11] They provided outlets and employment for writers, intellectuals, and novice politicians who considered themselves, with some justice, to be the elite of a new merit-based society and the makers of public opinion—at least, of the kind of public opinion that ought to matter. France possessed equivalents; like their British cousins, they usually had particular political perspectives, but not in every case. In all countries, most of the quality periodicals combined political commentary with articles on literature, art, and sometimes science and philosophy. They were essentially the places in which both "literature" as a concept and "literary criticism" as a practice were defined for the modern world. At the second level of importance were newspapers—not all newspapers, but the ones that, in each country, maintained a readership made up of the kind of people noted earlier. The *Times* of London was the English archetype. The editors of these magazines and newspapers typically had very close connections to both the political and the literary elites of their countries.

Voluntary associations constituted a much more significant element of the nineteenth-century European public spheres than they had in the eighteenth century (or than they do today, for that matter). These ranged from the famous organizations that aimed at effecting major changes in national policy (in Britain, the Anti-Corn Law League and the antislavery societies, which were models for similar associations in other countries) to special interest groups and lobbies.[12] They included scientific societies that facilitated exchanges among professional scientists (of whom there were relatively few early in the nineteenth century) and amateurs (of whom there were a great many). These societies essentially defined the scientific and social scientific disciplines that appeared in the nineteenth century and that became, in the form of departments, the structural basis of academia, but originally they were not conceived of as organizations of academics or even of experts.[13] They instead presented themselves as elements of the public sphere, speaking not just to specialists but also to the general, educated public on matters of importance, supposedly outside the arena of partisan conflict. Still other associations were devoted to "humanitarian" concerns and, from midcentury, to social issues. Most of the well-known issues of the mid- and late nineteenth century were essentially creations of the public sphere, originating through the interaction of societies and the media and then passing on to government. Among these was the "Social Question": what to do about the problems of a working class produced and exploited by capitalism and industrialization?[14] As we will see later, educational reform (including university reform) was the subject of a similar set of issues.

The quality media and the associations made up most of the structural element of the core public spheres of Western Europe. There was also a distinct cultural element, comprising informal understandings about how the conversations of the public sphere were to be carried out and more formal behavioral patterns that were themselves subjects of discussion. One of the most important of the understandings had to do with the practice that Habermas emphasizes in describing the eighteenth-century public sphere and which is often called "bracketing." "Bracketing" in the eighteenth century was an implicit agreement that, in particular places such as coffeehouses or clubs and in certain media, differences in social standing were to be temporarily laid aside in order that public discussion might take place freely and efficiently. The agreement had its limits: it did not apply to people without the education to contribute to the discussion or to people who lacked the manners to take part in it civilly. The bracketed public sphere thus implicitly included people who could be considered "gentlemen" in a behavioral as well as a hereditary sense, and extended perhaps a little more broadly to people who were "respectable."[15] In the European public spheres of the nineteenth century, the function of bracketing remained important, but it became considerably more complex, at least in terms of the ways in which it was envisioned. Some (far from all) of the active participants in public discourse began to describe themselves as standing *outside* the conventional framework of social classes, constituting what journalists came to call the "Fourth Estate." This went beyond an implicit agreement to lay social distinctions aside. It constituted an assertion that, regardless of their backgrounds, the people who wrote for quality periodicals comprised a category separate from the major social classes. Mainly on the basis of their educations and their connections with the media, they could claim to occupy a privileged point of observation from which they could offer "disinterested" and "objective" views of politics, society, and culture. By the late nineteenth century, this claim was transformed into a representation of the public sphere as a space *defined* by objectivity.[16] Not everyone accepted this representation, but it was widely held. Objectivity in this space could be achieved in two closely related ways: by adhering to a code of behavior that increasingly came to be identified with *professionalism,* and by adopting the practices of *science*—which, as we will see shortly, strongly affected the relationship between the public sphere and higher education.

One of the behavioral patterns universally accepted as normal and required (although not always practiced) in the conversations of the public sphere was *civility.* This was a holdover from the eighteenth-century public sphere, an adaptation to the purposes of public discourse of a "gentlemanly" way of acting. The essence of civility was the notion that the kind of interaction among people most conducive to order and the rational conduct of social intercourse is based on restraint of passion, manifest respect for the opinions and interests of other people (or at least of those who themselves display the same sort of respect), avoidance of both excessive familiarity and excessive formality, modesty without self-effacement, and deference without servility. The editors of the *Spectator* and other early eighteenth-century writers urged it specifically as a model for both oral and written controversies in public discourse.[17] Nineteenth-century versions of civility favored rational, restrained argumentation and perhaps irony and wit over extravagant denunciation and name-calling. Its practice was supposedly one way in which the quality media could be distinguished from others. Of course, reality often fell short of the ideal, but even so, someone writing for quality periodicals who did not maintain a civil tone ran a great risk of not being published, and if published, of not being accounted respectable and thus not attended to.[18] Increasingly in the nineteenth century, the ability to engage in civil argumentation was taken to be a sign not so much of hereditary gentility as of having been *educated* properly to behave like a gentleman—for the most part, at a university.

Certain modes of exposition were central parts of the culture of the nineteenth-century public sphere. History, as both a literary form and a way of building an argument, was extremely popular. The major positions on politics and ideology were presented, validated, and elaborated in historical works by politician-historians such as Macaulay and François Guizot.[19] History permitted writers to bring the practices of moral judgment and objective observation and analysis together into a seemingly coherent conceptual enterprise. One of the claims for the efficacy of open, public discussion—that is, for a free public sphere—was supposedly that it facilitated objectivity and correct moral judgment simultaneously.[20] In open discussion, values could be weighed and articulated, means to attain moral ends could be compared, people could attempt to persuade each other of the truth or utility of a particular set of ideas, and the grounds for persuasion could be analyzed for all to see. Critical history, based at least to some extent on primary research and written in either essay form or as an extended narrative that incorporated interpretation and judgment, was a means of undertaking this kind of discussion.

Writers conducted the published conversations of the core public sphere in a wide variety of formats: frequently in straightforward works of history, sometimes in treatises (particularly in the sciences and the social sciences), and universally in essays and book reviews appropriate for the quality periodicals. Published sermons were still very important, although their popularity waned markedly at the end of the century. And of course, there were novels. The nineteenth century was the highpoint of the novel as a medium for intervention in the public sphere. Not only had some novelists learned to conceive of what they were doing as a high cultural activity that aimed at producing "literature" (i.e., fictional writing worthy to be reviewed in the quality periodicals), but a great many of them had molded the genre to make it a means of influencing public opinion on particular issues: Dickens did so in his own peculiar way; so did Emile Zola in France and Gustav Freytag, the author of Germany's runaway midcentury best seller *Soll und Haben*. And so did Harriet Beecher Stowe, whose *Uncle Tom's Cabin* was perhaps the best known of all nineteenth-century novels that aimed at making a case in the public sphere.

Uncle Tom's Cabin brings up three important topics: first, gender and the public sphere; second, the relationship between the kind of elite or core public sphere we have been discussing and the broader discursive public of the more popular media (Habermas's "mass" newspapers, magazines, and novels); and third, the Atlantic aspect of the public sphere. Although most of the active contributors to the core public spheres of the nineteenth century were men, and relatively few women occupied salaried positions in them, the public sphere was generally open to people of both sexes. This was true not only of woman novelists, who abounded in the mid-nineteenth century at all levels of popularity, but also of woman essayists. George Eliot had a highly successful career as both, and also as an editor, and she was far from alone. The core public sphere in most countries was, in this regard, considerably in advance of trends in other areas of public and professional life.

Clearly, although Stowe was heavily involved in public discourse before and especially after the publication of *Uncle Tom's Cabin* in 1852, her novel was not aimed primarily at, say, the readers of the *Edinburgh Review* or even of the *New York Evening Post*. Yet it had a distinct political and ideological message, which Stowe spelled out and which she had her characters enact. She was performing in all areas of the public sphere, which resulted in a work that was discussed in all the major reviews in America and Europe and by large numbers of readers of varied social standing. Her principal stated target was intelligent American Southerners (who despised the novel), but her actual audience was the totality of literate, respectable people: the same people who read Dickens's novels, the same people who were the audience for the large-circulation press made possible by the mid-nineteenth-century technological revolution in printing. They were mostly, but not exclusively, people in the middle class.[21]

This audience and the media that appealed to it did not destroy the elite public sphere in Europe, but did require substantial adjustment on the part of its central personnel. For one thing, particularly in Britain, it was increasingly necessary to affect the thinking and political behavior of respectable middle-class people because, as more of them received the vote and as more still took an interest in politics, it was they who became the reference point for political life. It was not that governmental leadership left the hands of elite groups (although those groups underwent substantial changes), but national leaders had to speak, coherently, to a wider segment of the population than the normal readership of the *Edinburgh Review* or even the *Times* of London. For another, the commercial possibilities of a large-circulation press, for both making money from subscriptions and carrying advertising, were impossible for publishers to avoid. At midcentury, there was an explosion of new periodicals: not just, or even primarily, of the "scandal sheet" variety that became a principal object of criticism within the American core public sphere toward the end of the century, but of the "middle-brow" type, which included short, informative articles on subjects of serious import and increasingly large amounts of illustration. The best known of these was the *Illustrated London News*, which quickly elicited an American imitator. Older newspapers adopted new formats, and new newspapers appeared to cater to a broader audience.[22] This was part of the context in which *Uncle Tom's Cabin* made its impact. Clearly, in order to work in this context, a rethinking of the structure and mode of the elite public sphere was necessary. People who perceived this constructed a model of a "public" consisting of two parts: an inner, intellectually exclusive sphere in which the quality periodicals operated, surrounded by a larger outer sphere that levied a less onerous tax on the mental and educational capacities of its readers. The latter received issues, positions, and intellectual consensus (when it existed) from the former. The model did not replicate the complexities of reality particularly well, but it was very influential when, after the Civil War, Americans set about constructing their own version of a modern public sphere.[23]

The other interesting point about Stowe was that she was an American. She was one of several American writers who obtained international reputations in the nineteenth century prior to the Civil War, along with Ralph Waldo Emerson, William Ellery Channing, Nathaniel Hawthorne, and (best known of all) James Fennimore Cooper. Apart from Cooper, she was the only one to obtain a really wide audience, and like Cooper's, her international audience lay in large part outside the elite public sphere. Nevertheless, the fact that her book was available and reviewed in Europe almost immediately after publication in the United States indicates the existence of a transatlantic connection: an Atlantic public sphere, the center of which, as far as the United States was concerned, lay in Britain.

The Atlantic Public Sphere and Pre-Civil War American Public Discourse

As both Americans and foreign visitors frequently noted before the middle of the nineteenth century, the United States did not possess its own version of an elite, modern public sphere. Rather, the United States occupied a peripheral position in an Atlantic public sphere, the center of which was the public spheres of the major Western European countries—preeminently Great Britain. What the people who created an American core public sphere after the Civil War were consciously trying to do was to change this situation, not by cutting off the Atlantic connection but by establishing an autonomous American element that could balance the British one. This intention was, as we will see, crucial for defining the connection between the core public sphere and the reconstruction of American higher education that took place during the last third of the nineteenth century.

Public discourse had been an essential part of American political life since before the beginning of the republic. Some scholars have seen developments akin to those in eighteenth-century Britain as the formation of an American public sphere in Habermas's sense.[24] In the explosion of hyperactive partisan politics in the 1790s, however, hundreds of local newspapers with party associations had appeared all over the country. This flood of journalism ebbed somewhat later, but it made the local, partisan newspaper a central feature of political life in the nineteenth century and swamped the remnants of the earlier public sphere. In the first years of the nineteenth century, circles of educated people with political and journalistic connections formed in the major cities. These urban intellectual elites did not, however, manage to retain anything like a central role in national political life, nor were their views, for the most part, consistent with the increasingly aggressive democracy that developed during Jefferson's administration and afterward. They tended to be Federalist in politics and therefore lost significance and self-confidence with the collapse of the Federalist Party after 1808. They were in no position to provide a center of private conversation around which a more structured, institutionalized, and above all *national* public sphere on the nineteenth-century European model could develop, nor were they able, despite frequent individual attempts to do so, to create high-quality journalism in America.[25] Even after more broadly based intellectual elites appeared in the 1820s and 1830s, their members functioned largely in local urban or state contexts; when they turned their attention toward major cultural or intellectual trends, they looked to Europe.

In a very real sense, the intellectual elites of the United States came to occupy a secondary and dependent, almost a colonial position with regard to the quality magazines and associations of the modernizing nineteenth-century British public sphere. The only lasting attempt before the 1850s to establish an American equivalent to the British reviews and quarterlies was the *North American Review* (founded in 1815), but its readership was never large and the journal itself was periodically moribund.[26] For the most part, American intellectuals who wanted to sample the life of a "real," moderately elitist public sphere had to do so at a distance, by subscribing to the *Edinburgh Review,* for example, or by reading reprints of European articles in digests such as *Living Age*. Although William Cullen Bryant edited the New York *Evening Post* with the intention of making it the newspaper of educated readers throughout the country, the structure of public discourse constituted by the *Post,* the *North American Review,* and the handful of other publications with national pretensions seemed— and was—puny compared with the rich and vibrant quality journalism that existed in the leading European capitals.[27]

In the United States, there was certainly a widespread desire to create a public discourse distinct from European ones—particularly a discourse that emphasized democracy.[28] On the other hand, many factors worked against the emergence of a full-fledged public sphere. The

United States was both politically and culturally diffuse, and its regions displayed substantial economic differences from each other. These circumstances helped to produce a peculiar situation: a political culture that was aggressively nationalist and democratic and that flaunted America's self-proclaimed superiority over the older nations of Europe, but at the same time a public discourse that was fragmented, localized, and sometimes incoherent. National public discourse had no clear center and, as far as discussion in periodicals was concerned, depended largely on Western Europe (especially Britain) to provide it with one. Some scholars have suggested that, before the Civil War, the United States was not really a modern nation-state at all, a least not one with a fully articulated, powerful central government.[29] It had no real capital city (Washington was a political capital only in a formal sense, while New York had not acquired its economic and cultural centrality), a fact that further encouraged the fragmentation of public discourse. The media that operated at the core of the British public sphere essentially served the same function for the United States, while the issues that were formulated and debated in Britain—issues that arose mainly from the confrontation of the British nation-state with social, economic, and cultural changes and with global political events—tended to inform the issues discussed in the United States.[30] Even positions on slavery and abolition, matters of very direct concern to the United States, became standard in America when they were formulated in European journals.

Foreign visitors to the United States in the second quarter of the nineteenth century consistently remarked on what they perceived to be a great deficiency in the public life of the country. To Frances Trollope in the late 1820s, the problem was not an insufficiency in the number of people interested in public affairs.[31] Quite the contrary: *everybody* (or everybody who was male) talked about politics; everybody read newspapers (which, however, were hardly worth reading) and supported a political party. Public life was active and, to Trollope, obnoxiously intrusive, making the actual range of tolerated views on any subject rather narrow because of the dominance of conventional popular opinion in a democratic society. What was absent (along with good manners, Trollope's central concern) was real *conversation:* intelligent discussion of a range of subjects into which considerations of politics could be inserted, among cultured people who actually knew how to converse and who did not instantly declare certain opinions and subjects to be off limits. There were, according to Trollope, a few such people in the United States, mostly in the upper classes of the major eastern cities, although she did not meet many of them and believed they had little general influence. The absence of enlightened conversations meant that the core elements of a true public life were absent. Until they arose, American society could not compete with its European counterparts in any arena except possibly the making of money.[32]

Another important critic, the French civil servant, historian, and political philosopher Alexis de Tocqueville, had more favorable things to say about aspects of American life than Trollope did, but he agreed with her about the weakness of public discourse. Unlike the French press of the 1830s, he said, the American press was unable to sustain an extended discussion of issues that were at all complex or abstract. "In America, three-quarters of the enormous sheet [by French standards, American newspapers were physically large] are filled with advertisements, and the remainder is frequently occupied by political intelligence or trivial anecdotes: it is only from time to time, that one finds a corner devoted to passionate discussions."[33] (By "passionate discussions," Tocqueville meant comprehensive expositions of complicated ideas to which the writer was deeply committed.) According to Tocqueville, Americans were far too concerned with commerce and with questions of short-term policy to be able to take the long view of anything. Taking the long view, he thought, was the natural function of an aristocracy. Without an aristocracy, the Americans needed to adopt an appropriate structure for supporting detached, objective thinking and "passionate" advocacy of political, literary, and philosophical ideas. He recommended for the United States the kind of intellectual and political elite that had formed around quality journals and newspapers in European capitals and that had created political parties committed to general political and philosophical principles—in other words, a modern public sphere. Many prominent American intellectuals agreed with him.[34]

As the United States faced its growing crises of slavery and sectionalism in the late 1840s and the 1850s, the absence of the kind of public sphere that existed in Europe was felt even more insistently than ever. During the same period, consciousness of rapid economic change and urbanization reinforced the feeling that there was a deficiency in American capacity to discuss such things. Observers of the international scene expected that the United States was going to have to play an expanded role in world affairs. They saw the lack of a coherent public sphere not only as a handicap when it came to making policy under such circumstances, but also as a symbol of American weakness in comparison with other powers, especially Britain and France.[35]

American publishers deliberately took a number of major steps in the 1850s to bring the United States up to European standards in public discourse. The decade saw the establishment of the great American monthly magazines: *Harper's* (1850), the *Atlantic Monthly* (1857), and the ultimately less-successful *Putnam's* (1853).[36] The writers and editors associated with the new monthlies consciously aimed to establish in the United States an equivalent to the public spheres of Britain and France. The first generation of American quality journals possessed distinctive orientations on some issues (e.g., slavery abolition), but not the ideological or party affiliations that their European counterparts had. They were also primarily commercial ventures whose owners expected them to make profits by identifying or developing an audience for articles that covered a broad array of "civilized" interests. Their book reviews, increasingly the products of first-rate writers and prominent intellectuals, quickly attained the standards of the best European publications. They published a great deal of fiction, thereby providing a platform for the so-called American Renaissance of the pre-Civil War years.[37] In addition, the 1840s and 1850s witnessed the development of a more extensive, higher-quality daily press, primarily in New York. By the end of the 1850s, the New York *Evening Post* was no longer the sole daily with meaningful literary aspirations and national resonance. The *New York Times* had begun its plodding but inexorable advance toward its twentieth-century role as national "newspaper of record."[38] Other New York dailies, such as Horace Greeley's *Tribune* and James Gordon Bennett's *Herald,* had readerships that extended broadly.

Despite the new periodicals, however, it was clear to observers that the United States still did not possess the kind of network of journals and writers linked to associations, politics, and government that existed at the center of the Atlantic public sphere in Great Britain. Nor was it really possible to make in the United States the kind of career that, for example, Macaulay had done in Britain. When Henry Adams attempted to do so in the late 1860s and early 1870s, the key to his strategy was to establish his reputation by publishing in leading *British* journals.[39] The American monthlies were too few in number and (apart from the fact that they tended to oppose slavery) too unwilling to be identified with particular positions in politics to create by themselves the basis for a lively public discourse. Contextual factors, especially localism, the lack of a political and cultural center, and the absence of a fully articulated modern nation-state, continued to work against the formation of an American equivalent to a European public sphere and to maintain the previous structure of the Atlantic sphere.

Then came the Civil War. Not only did the war produce a modern central state almost overnight, not only did the war and its aftermath create a host of problems and issues that cried out for discussion on a scale and with a sophistication almost unprecedented in America (or anywhere else), but the war also shaped the outlooks of a generation of educated Americans eager to respond to the situation. This "Civil War generation" is crucial to our story.[40] They were not the first Americans to try to construct an autonomous, modern public sphere in the United States, but they were the first to succeed. They were also the people who rebuilt American higher education and created the modern American university. The two achievements were closely connected.

Building the American Core Public Sphere

To young people of education and social standing who had lived through the Civil War—some of them as officers in the armies—the belief that the United States could never be the same as it had been before 1861 was a commonplace that they expressed in various ways. Oliver Wendell Holmes Jr. thought that his experience of combat had given him a glimpse of reality that exploded the idealistic myths of his youth and made him utterly suspicious of dogmatic convictions and ideological beliefs.[41] Henry Adams, who had been neither in the army nor in the country during most of the war, later reported that what he primarily noticed on his return in 1868 were manifestations of new forms of mechanical (and by extension, social) power and his own lack of understanding of them.[42] Other members of the Civil War generation of educated Americans emphasized different aspects of the changes that had come about because of the war. One thing they almost all noted, however, was the expansion that had taken place in the scope and authority of the national government. The United States, they recognized, had become a modern nation-state.[43]

For the most part, the leading figures in the Civil War generation expressed a profound ambivalence toward a new political world of full and apparently unbridled national power. In the past, such a world had been thought to exist only in Europe, and most Americans, even those who admired Western Europe, had believed the United States lucky to have avoided it. The Civil War had shown them that it could no longer do so.

And yet, there were good reasons for their distrust. Who knew when the assumption of essentially dictatorial power by the national government in wartime would be repeated under other circumstances? Who could tell whether the powers of the federal government would fall entirely into the hands of corrupt and unworthy politicians thrown forward by the twin forces of democratic machine politics and expanding big business—as seemed to be happening under the Grant administration?[44] At the same time, not only was the new nation-state, the modern Leviathan, apparently necessary and inevitable, but it potentially offered the means for effecting desirable changes in society, especially in dealing with the massive dislocations created by industrialization and urbanization and in solving problems arising from immigration and the abolition of slavery. How was it possible to obtain the advantages afforded by a modern nation-state while avoiding as many as possible of its dangers? How could you ride Leviathan without tempting Leviathan to swallow you? Various answers to these questions suggested themselves to members of the Civil War generation (usually including civil service reform), but almost everyone agreed that what was vitally needed was an array of connected publications and associations in America—not in England—in which the possible answers could be proposed, examined, and debated intelligently. The array had to be sufficiently well connected to centers of political power that the deliberations would have some effect, and yet sufficiently independent to avoid "corruption."[45]

Then there was the matter of a career. Many college educated young men after the Civil War found that their plans for their lives had changed, whether because of alterations in the larger society or because their experiences had redirected them. Young women with education looked for respectable careers in much larger numbers than ever in the past. Both before and after the war, many educated young men from Europe moved to the United States seeking to take advantage of opportunities offered by an expanding economy. In some cases, these people had expertise in specific technical fields. In others, they had ambition, inclination, or access to capital that brought them directly into business. But many of them possessed principally the knowledge and skills that were the products of a liberal education, whether in an American college (or seminary for women) or in a European classical secondary school or university. What were they to do with these things? In Europe, the answer would have been obvious: a learned profession (especially the law), the civil service (closely linked to the law in France and Germany), or the public sphere (possibly going from there into politics). The learned professions certainly existed in America (although critics claimed, with considerable justice, that most of their members were not all that learned),[46] but not a European-style civil service or, as we have seen, a European-style core public sphere. It is therefore not surprising that members of the Civil War generation set out to establish the last two and to raise the standards of the first. Professionalism, civil service reform, and the articulation of the core public sphere went hand in hand from the 1860s through the 1880s. All three depended fundamentally on changing the nature of American higher education.

The people who accomplished these things by no means agreed about all the issues of the time, nor did they come from exactly the same sorts of background, but they resembled each other sufficiently that they can be thought of as a group. Also, many of them knew each other—in many cases right from the 1860s, and if not, then in the course of establishing the networks of the new public sphere. The brothers Henry and Charles Francis Adams can serve as examples. (The fact that they were sons of the Civil War American minister to Britain and grandsons and great-grandsons of presidents makes them less than completely typical, but otherwise they fit the bill reasonably well.) After graduating from Harvard (and, in Henry's case, a period of study in Germany), they had been caught up in the Civil War. Charles served as an officer in the Union Army, while Henry worked as his father's secretary in London (acquiring an intimate knowledge of the British public sphere and getting some pieces published in the process). In his autobiography, Henry Adams recounts discussions with his brother in 1868 about what to do with one's life when one's education and prior experience provided no usable knowledge of "modern" subjects such as technology or the economics of industry. Charles Adams decided to learn about railroads and to make his future writing about transportation issues. Henry turned to journalism with a focus on monetary policy, which also entailed learning about things outside the curriculum of formal classical education.[47] The Adams brothers wrote pioneering exposés of corruption on Wall Street and on the railways (one of the most important of them published in London).[48] Charles went on to become chairman of the board of the Union Pacific railway, while Henry, having decided that he could not yet emulate Macaulay in America, accepted a position as assistant professor of medieval history at Harvard—a job that brought with it the editorship of the *North American Review*.[49]

Henry and Charles Francis Adams were therefore in at the birth of the American core public sphere, and both were entirely aware of what they were doing.[50] A host of others went through similar patterns, acting as writers and journalists and editors in the quality press before turning to other occupations. Many of them continued to contribute to magazines for the rest of their lives—writing not only about what turned out to be their professional specialties, but across the range of topics of public importance in the late nineteenth century. Their exchanges in the quality press and at the meetings of the associations they founded quickly came to constitute, as intended, an active, intelligent, and highly informed public discussion conducted by an educated elite before a wide audience. Some of those who took part in the discussion included (besides the Adams brothers) George William Curtis (lead editorial writer of *Harper's*), Frederick Law Olmsted, Charles Eliot Norton, James Bryce (a Briton, but the most respected interpreter of Gilded Age Americans to themselves), William James and many of his associates in Cambridge, and most of the leading figures in the reconstruction of higher education (especially Charles W. Eliot and Daniel Coit Gilman).[51] They were followed by a younger generation, several of whom—for example, Theodore Roosevelt and Henry Cabot Lodge—were able to follow careers in what was by the 1880s a well-established core public sphere and ended up, fairly quickly, in politics.

Media

Quality magazines and associations were the central structural features of the new core public sphere. The magazines included the general interest monthlies established in the 1850s (the *Atlantic Monthly, Harper's,* and—for a while—*Putnam's*) and several others (the *Century* and *Galaxy* among them) created in their image. But the most active, deliberate agents at work constructing the core public sphere were the weeklies, preeminently the *Nation*.

The *Nation* was established by a group of prominent radical Republicans in 1865 as a weekly journal dedicated to projecting the cause of abolitionism into the era of Reconstruction. Very quickly, however, its nature and purpose changed, mainly because E. L. Godkin, its first editor, had other plans for it.[52] Godkin consciously set out to create a periodical that would continuously operate at the center of public discussion of matters of importance, with a decided slant on those issues that would govern the individual positions taken by the journal's writers. It would cover a range of topics, including literature and art, but its emphasis would be primarily on politics. Godkin, an Anglo-Irishman who started his journalistic career in Belfast and moved to the United States in the

1850s, probably had an explicit British model in mind: the *Economist*. The *Economist*'s editor, Walter Bagehot, had constructed for himself a place in the process of British policymaking and public opinion formation that Godkin wanted to occupy in America.[53] Godkin never quite made it, but he accomplished a great deal anyway.

Godkin began the *Nation* in close cooperation with a number of relatively like-minded people, including Olmsted and Norton. Olmsted had first made his reputation in the 1850s as a journalist reporting on the South, while Norton, who had also been a traveling journalist in the 1850s writing prominently for the *Atlantic Monthly,* had been the editor of a unionist periodical during the Civil War and was, at the time of the *Nation*'s founding, coeditor of the *North American Review.* Olmsted went on to be the dean of American landscape architecture, while Norton became the first professor of fine arts at Harvard. Godkin used his contacts with these and other associates to develop a large (eventually immense) connection of reviewers and article writers among educated people of his generation—a connection whose membership reads like a directory of the intellectual and political elite of late nineteenth-century America. Many of them continued to contribute to the *Nation* for years, although because Godkin possessed an overbearing personality and insisted on imposing his own point of view on articles that touched his fundamental concerns, many parted ways with him as time went on. This was true, for example, of Charles Francis Adams and Henry Adams, both of whom contributed important pieces in the 1860s and 1870s but not much later. Charles Eliot wrote for the *Nation* and corresponded with Godkin before becoming president of Harvard in 1869, and he continued to do so afterward. Daniel Coit Gilman was another frequent contributor. A significant part of the reputation that bore him from a junior teaching position at Yale to the founding presidencies first of the University of California and then of Johns Hopkins came from his articles in the quality magazines. When Gilman was building Johns Hopkins in the late 1870s and the 1880s as the first avowedly research-oriented American university, he paid close attention to the publications in such magazines of potential candidates for his faculty.[54]

Godkin thus managed to create not just a significant periodical that dealt in depth with matters of public interest, but also a network that constituted an important part of the new American public sphere and extended to Europe. Of Godkin's many British contributors, the one to whom he and many of the other founders of the core public sphere stood most closely was James Bryce, later Viscount Bryce. Bryce illustrates the connection that continued to exist between the newly constructed American core public sphere and its European models. He was also one of the first Europeans to recognize that the connection across the Atlantic had suddenly become one of equals. Bryce was a multitalented man—a distinguished historian, lawyer, Liberal politician (briefly a cabinet minister), and at the end of a very long career, one of the most influential of all British ambassadors to Washington. In the 1870s, he began a series of extensive visits to the United States that made him a familiar figure in the American core public sphere. The principal literary result of Bryce's American connection was a book entitled *The American Commonwealth* (1888), which remained for nearly half a century the standard explication of American democracy. *The American Commonwealth* closely reflects the "Mugwump" political perspective of Godkin and his associates.[55] In its third volume, it contains the first comprehensive treatment of the formation and the nature of the core American public sphere.[56]

Bryce participated in other ways in constructing and publicizing the core public sphere. Together with others, many of them connected with the *Nation,* he helped to organize some of the special purpose disciplinary societies that formed another significant segment of the American public sphere as well as a central factor in American academia. He served a term as president of the American Political Science Association. Bryce was also among the first to collect information in depth on American higher education during its period of expansion and definition in the late nineteenth century and to point out many of its implications for democracy and public life.[57]

Godkin's *Nation* illustrates many of the salient characteristics of the media side of the core public sphere in its first years—and in many respects, for decades after. The founders of the magazine consciously viewed it as a new and vital element of American society, which was intended to emulate and equal what had until that time been the British center of the Atlantic public sphere without eliminating the Atlantic element altogether. The *Nation*'s operations depended on a network of contributors who were educated people in various elite walks of life. They were mostly men; unlike the *Atlantic Monthly,* the *Nation* seems to have published relatively few pieces by women, although because its articles were mostly unsigned in the nineteenth century, it is difficult to determine what the proportion was. Godkin and the contributors envisioned a readership that was educated, respectable, and responsible. Unlike the more general quality magazines or the *North American Review,* which tried to cover a range of positions on matters of public interest,[58] the *Nation* generally took one position: Godkin's. Godkin saw himself as an apostle of a fairly radical variety of free trade, free market liberalism and made sure to choose contributors who agreed with him. He took stands on most issues. Those having to do with, for example, civil service reform he shared with a wide array of members of the Civil War generation. On others he had less support. This characteristic of the *Nation* quickly revealed another: the financial insecurity of periodicals that took decided stands on issues. The *Nation* lost much of its initial readership when Godkin, a Republican with a mostly Republican audience, supported Democrat Samuel Tilden in the dispute over electoral votes that followed the presidential election of 1876. Thereafter, the *Nation* operated at best on the margins of solvency. It was bought in 1881 by the German American financier Henry Villard, who had just purchased the New York *Evening Post*, to be a weekly magazine for the *Post*.[59]

Villard's connection to the *Nation* tells us more about the core public sphere. It was not a simple case of a sellout to big business. Villard let Godkin direct the *Nation* just as before and made Godkin editor of the *Post* as well in 1883. It was Villard's son, Oswald Garrison Villard, who eventually began to run the *Nation* and who took over its formal editorship after Godkin's death in 1902, bringing the magazine much farther (by Godkin's standards) to the left. Henry Villard, who immigrated to the United States from Germany in the 1850s, actually started out as a journalist working for German American newspapers.[60] He became an ardent abolitionist (he married William Lloyd Garrison's daughter) and an active member of the group of people who, in the 1860s, organized the civil service reform movement, established the American Social Science Association (ASSA), and started the *Nation*. Villard had hoped to make a career somewhere within this context—a career, in other words, in the public sphere, built around his paid position as secretary of the ASSA.[61] His plan did not work out, however, so he used contacts he had made as a journalist to go into business as an agent for German firms wanting to invest in the United States. On that basis, Villard built up a complex empire of railroad holdings, organized the Northern Pacific railway, and bought the *Post* and the *Nation*. Again using German contacts and capital, he later reorganized Thomas Edison's companies into Edison General Electric.

Villard is interesting in terms of our topic for several reasons. He participated in the creation of the core public sphere, not just as an "angel" for the *Nation* but even before he went into business, as a journalist and a leading member of several movements and organizations. After his move into commerce he did not involve himself heavily in politics, but he retained a progressive liberal outlook and an interest in issues of social change. He was a thoroughly Atlantic person, with German contacts not only in business but also with the parties and journals of the German liberal left.[62] Thus, although Villard did not write much after he gave up journalism, he was still an active and important figure in the core public sphere. His involvement gives evidence of the importance that it had so quickly achieved in the United States and of how it worked. He was also the principal force behind the creation of the University of Oregon, partly in pursuit of the interests of the Northern Pacific Railroad but partly also because he, like almost all the other founders of the core public sphere, believed that an expanded and reformed system of higher education was essential in the United States.

Godkin was not the only person to organize a network of contributors and acquaintances and to try to use the network to influence policy, although his network was probably the most extensive of the late nineteenth century. Nor was the *Nation* the only significant periodical in the new public sphere, although it was in many ways an archetype. The large-circulation monthlies continued to be major fixtures—or at least those, such as the *Atlantic Monthly* and *Harper's,* that managed to survive to the end of the century. Other magazines of a more specific ideological character or with more particular subject orientations also proliferated, although they, too, had a high death rate. The *New York Times* challenged and then replaced the *Evening Post* as the principal national daily newspaper of the core public sphere.

The people who operated in the new public sphere had a fairly straightforward idea of what its structure should be and how it should operate—an idea heavily influenced by European models. The public sphere was to function on a national level. Its key feature was to be the quality periodicals, which would be closely tied, through networks of contributors and their personal contacts, to other elements of the public sphere—to politicians, to businesspeople of Villard's sort, to the universities, and to what reformers hoped would be a modern federal civil service staffed at the upper end by people like themselves.[63] What was discussed at the core, in the quality periodicals, would in turn affect what was published in the broader-circulation newspapers.[64] (The relationship between the *Nation* and the *Post* was in some sense supposed to model this process of broadening the public sphere, although the *Post* reached only a fairly restricted audience.) The conversations that followed this pattern would be the primary means for achieving intelligent consensus about solving the problems of state power and social and economic change in the United States, and also for raising the level of cultural awareness in America to a par with that of Europe.

In fact, the core public sphere never worked entirely that way, despite its spectacular success in shaping the meaning of reform in late nineteenth- and early twentieth-century America. For one thing, large-circulation, popular newspapers like those of James Gordon Bennett, Joseph Pulitzer, and eventually William Randolph Hearst were not always content to let their "betters" take the lead. They showed a distressing tendency to create issues of their own, and they tended to be notably more solvent than the qualities.[65] For another, there was a substantial increase in the number and the circulation of "middle-brow" periodicals, aimed at a much wider audience than the magazines of the narrow core public sphere but edited with the intention of addressing people of some education and intelligence. *McClure's Magazine* was one of the models of this type. Such magazines became significant centers of discussion in their own right.[66] Nevertheless, if we expand the notion of the quality press to include periodicals like *McClure's*, it is clear that well before the end of the nineteenth century, it constituted one of the most influential factors in American public life.

Associations

One of the ways in which the Civil War generation consciously set out to parallel developments in Europe was by establishing voluntary associations that focused on significant, but limited, ranges of knowledge useful for public debate and state action. "Parallel" is the correct word. This was part of an Atlantic phenomenon. On both continents, the period between 1850 and the early twentieth century saw a proliferation of intellectual associations and of organizations that were supposed to bring science to bear on solving social problems before an educated, intelligent audience. The movement began somewhat earlier in Europe, but the United States and other American countries caught up rapidly, creating a lively transatlantic exchange of models and ideas.[67] Occasionally, the Americans were ahead of the game. The ASSA, founded in 1865, preceded its vastly more famous and successful German analog, the *Verein für Sozialpolitik*, by seven years. Some of these new associations eventually became mainstays of organized academic disciplines, but as in Europe, they were not originally envisioned in that way. Initially, their founders thought of them as independent vehicles for discussion and research that would strengthen the public conversation and prevent it from being overwhelmed by the forces that threatened it.

Associations devoted to intellectual subjects and public issues were not a novelty in the United States. In the period after the American Revolution and at intervals throughout the first half of the nineteenth century, attempts had been made in practically all American cities to establish such societies. The few that had survived had generally evolved into fixtures of local urban elites, clubs that attracted membership from the wealthy and distinguished and that showed a persistent tendency toward dilettantism. Nevertheless, some of them did useful work. They contributed to the formation of urban cultures. Many of them concerned themselves with social issues; with encouraging scientific "progress"; and with creating, expanding, or reforming educational institutions. As Thomas Bender has shown with regard to New York City, some of the older, elite-dominated associations gave way as time went on to more open ones that took on the characteristics of serious scientific and cultural societies resembling European models.[68]

Despite these beginnings, it was the period after the end of the Civil War that witnessed the formation of the *national* organizations that, to this day, constitute a significant part of the structural basis of both the core public sphere and academia in America. One of the first efforts was the creation of the ASSA, which was supposed to be a grand organization that brought scholars and researchers together with journalists, businesspeople, politicians, reformers, and officials to discuss social problems and to coordinate the mobilization of social science for their solution. Its list of founders looks as though it were drawn from the *Nation*'s pool of contributors and financial supporters.[69] But although the ASSA attracted a great deal of media attention, it never achieved very much. Some of the reasons that it did not are important for understanding the nature and limits of the core public sphere and its relationship to higher education.

In the first place, both the ASSA's membership and its envisioned scope of activities were extremely broad, which meant that its organizers found it difficult to identify a consistent focus—unlike, for example, the various associations favoring civil service reform that shared many of the same leading members. Moreover, because of the association's breadth, a great many of the subjects that were discussed at its meetings were controversial and threatened splits in the membership. Because most of the members were essentially there out of an interest in participating in public discourse rather than because they were experts in any field of social science, the association was seldom able to identify and carry through a program of research on particular issues. This last point constituted a major difference between the ASSA and its German cousin, the *Verein für Sozialpolitik,* which was similarly constituted but was dominated from the start by academic experts in social science fields (a minority of the membership) who were able to mount significant research efforts.[70] In other words, although the ASSA (like the *Verein*) was conceived as a structural manifestation of the public sphere—an entity that could provide space, occasion, and support for an appropriately bracketed conversation among intelligent, educated people on matters of general importance—it failed in large part because that did not seem to be sufficient. What was needed, so it was thought, was a greater emphasis on expertise and more narrowly focused topics.

The perceived failure of the ASSA was a major factor leading to the subdividing of the general field of scientific engagement with society in the United States into separate disciplines, the definitions of which generally originated in Europe. Starting in the 1880s, disciplinary associations appeared in rapid succession, reflecting to some extent the construction of professions in the fields they covered and the formation of academic departments at the leading American universities. The first was the American Historical Association (AHA) in 1884, then the American Economic Association in 1885, followed after an interval by organizations in newer fields: the American Anthropological Association (1902), the American Political Science Association (1904), and the American Sociological Society (1905). Still, the founders of most of these organizations saw them not primarily as professional or academic societies, but rather as straightforward adjuncts to the public sphere that permitted disciplinary expertise to be brought to bear on significant issues.

To take one example, the AHA initially reflected a vision of historiography not primarily as a university discipline but as a central activity of public discourse and of politics. There were only a few professors of history in the United States at the time of its founding; most serious historians were still self-supported gentlemen (like Henry Adams after his teaching stint at Harvard) or intellectuals trying to get into politics (like Theodore Roosevelt). Both Adams and Roosevelt served terms as president of the AHA. The first generation of Americans with graduate degrees who became professional historians did not appear until almost the 1890s. Within a few years, however, they and academic immigrants from Europe had transformed the historical profession and participated in a major realignment of history's connection to the public sphere.[71] The AHA reflected this realignment, although it continued (and continues to this day) to take its nonacademic functions quite seriously. In 1899, for example, at the height of a conflict over American imperialism in Cuba and the Philippines, the core public sphere was split down the middle.[72] The AHA decided to intervene, not by trying to settle the issue itself but by providing a historical and comparative perspective on a matter of policy flowing from the issue. The association commissioned A. Lawrence Lowell (one of the founders of the discipline of political science in the United States and later president of Harvard) to prepare a study based on two assumptions: the United States would hold on to the Philippines and thus be in possession of a genuine overseas empire, and the United States would establish a colonial civil service in order to administer the empire.[73] How should the colonial service be organized? Lowell's study compared the ways in which Britain, France, and the Netherlands selected and educated the members of their colonial services. He recommended (among other things) that the United States adopt the British model of recruiting the upper level of the colonial service directly from the elite universities. The study was presented in summary to the AHA at its annual meeting in December 1899 and published as a book the next year. In this case, the AHA saw itself as a neutral public resource for identifying a problem and providing recommendations prepared by an expert. The recommendations were delivered at a public session of the association's annual meeting, discussed by members of the association as representatives not only of their profession but also of the educated public, and then published for broader public discussion (not, it might be noted, delivered directly to the government). The association was, in other words, acting out a particular conception of the way an organization within the core public sphere should operate.

The disciplinary associations were not the only organizations in the core public sphere of the late nineteenth century. Many of the same people who contributed to the quality magazines, who established those magazines, and who tried to make careers in or through the public sphere also organized a wide range of societies devoted to accomplishing particular purposes, most of them cast as *reforms:* the National Civil Service League, which included almost everybody who was anybody in the core public sphere in the 1880s; the Anti-Imperialist League, formed in 1898, which represented one (large) fragment of elite public opinion; and a host of others.[74] In fact, "reform" itself, as a concept and as a way of visualizing purposeful change in government and society, was to a considerable extent a construction of the core public sphere. The term "reform"—orderly, nonrevolutionary, rational improvement—entered political vocabulary in the United States largely through the Atlantic public sphere, having become a significant feature of British public life in the first half of the nineteenth century. Reform suited both the general outlooks of the founders of the core American public sphere and the logic of the public sphere itself. Reform was also the central feature of Progressivism.

The core public sphere essentially created Progressivism at the same time that it created itself. It did so through its founders' choice of comprehensive reform as the primary objective to be sought through the new structure of national discourse and through the links between the core public sphere and the popular media, politics, science, higher education, and the state.[75] Without the unifying concept of reform, the public sphere would have been difficult to build; without the public sphere, Progressivism would have been impossible. The essence of the movement lay in the national discussion of what was wrong with America that was conducted in the quality press and, increasingly, by nonpartisan associations, and in the formulation by the same means of proposals for what to do about it. Progressives did not share the same ideas in detail or even the same political outlook (and certainly not the same political parties). What they did share was the general attitude toward discussion and reform that was built into the new core public sphere, in which practically all the leading Progressives were very active participants.

Careers and Professionalism

As we saw, a great many of the founders of the American core public sphere conceived of it as a context within which an educated, intelligent person could make a career. On the whole, most of them also accorded enormous respect to professionalism. These characteristics are easy for us to misunderstand, and in the reasons that we misunderstand them lie the roots of some of the long-term problems of both the core public sphere and higher education in the United States.

When people like the Adams brothers or Henry Villard or even E. L. Godkin talked about making a career as journalists (in what we have been calling the core public sphere), they did not mean that they intended to become professional journalists in a twentieth-century sense. What they had in mind was the kind of career followed by Macaulay, Bryce, and others in Britain, which involved writing for magazines and newspapers and perhaps even working for them for a limited time (or for a long time, if one became an editor), but in which the essential movement was toward a position in politics or governmental service, in respectable business, or in some more informal situation of influence. From Europe, they had derived the idea of a civil service, the upper ranges of which were to be filled with people appointed because of education and promoted through merit, who would combine an official career with writing for the quality press, taking part in the activities of appropriate organizations, and perhaps engaging in scholarship. Matthew Arnold provided an example. Many American civil service reformers wanted something similar for the United States.[76] However, things worked out differently. The kind of civil service that was created by the Pendleton Act in 1883 and subsequent developments was not congenial to the European scholar-bureaucrat-pundit model. The primary means by which intellectuals whose careers lay in the public sphere got into office turned out to be through party politics, not employment in the civil service—by election or by appointment to a non-civil service executive branch position after an election (as when Theodore Roosevelt became a civil service commissioner in 1889 and assistant secretary of the Navy in 1897). The kind of civil service that emerged (to which most reformers reconciled themselves fairly quickly) was a *professional* service at all levels—in the sense that incumbents were expected to become experts in the specific, narrow range of operations they were paid to perform and to focus all their compensated time on performing them. Although the founders of the core public sphere had expected this kind of thing for the middle and lower levels of the reformed civil service, it was not what they originally had in mind for the upper, highly educated level that they hoped, following European models, would develop in the United States and that would be part of the core public sphere. For this part of the civil service, they envisioned a set of practices and lifestyles that we would today probably identify as "gentlemanly" rather than "professional" in that it implied a substantial amount of leisure to participate in public discourse (although that did not mean that incumbents would not work hard overall).

And yet at the same time, many of them expressed profound admiration for professionalism.[77] By this they meant, initially at least, a number of things. One was expertise, but not primarily in the context of a single-minded, lifetime focus on a particular occupation. Rather, they meant an attitude: a moral commitment to becoming very good at whatever it was one chose to do—which might be an established "profession" such as law from which one derived a living but might also include any of a range of other forms of activity, several of which one might engage in simultaneously. A person acquired such expertise in some part for self-satisfaction, in larger part for the public good, but most of all because realizing a moral commitment in this way manifested one's worth as an individual. Some members of the Civil War generation who had been army officers expressed pride in their own and their units' growing professionalism as the war went on even as they became disillusioned with the causes for which they were fighting.[78] It was the commitment to doing well what had to be done even under trying circumstances that mattered, not so much (up to a point) the end for which one was doing it. Civilization and social order depended on enough people showing the same kind of professionalism.

For another thing, professionalism was something that was developed, not inborn. The ability to become a professional was generally a result of education of a particular kind: one that promoted character, not by punishment or by teaching abstract classical examples of virtue, but by placing individuals in situations in which it was necessary to acquire knowledge by their own efforts (not by rote memorization), and necessary also to make free choices and be responsible for the results. (To anticipate slightly, one of the prime objectives of the post-Civil War reform in American colleges and universities was to create this kind of education.)[79]

Finally, professionalism was a *public* pattern of behavior, not just in the sense that the public was supposed to be a beneficiary of what people acting professionally did, but also because professional activities were carried out in a public context. Standards for professional activity were publicly advertised, whether for lawyers and physicians or for, say, somebody writing an article for the *Nation,* and professionals were at least morally (and sometimes legally) responsible to the public for adhering to them. One of the principal goals of a professional was supposed to be to achieve public respect on account of professionalism, which counted for more than making a great deal of money.

Most important for present purposes, professionalism meant participating in the public sphere. We still have the same expectation, except that we usually conceptualize it in the limited context of the technical practice of a regulated profession such as medicine. A practitioner is supposed to read the journals in his or her field and to write up discoveries or observations for other practitioners' use. This was true also in the nineteenth century, but the expectation was broader: a professional was supposed to communicate ideas and take part in public discourse on a wide range of issues, not just matters of specific technical expertise. If insights from that expertise were applicable to what was being discussed, so much the better, but the crucial point was to contribute in a way that could be understood by an educated public—which required that professionals possess a kind of education that allowed them to do so. This was a central feature of what it meant to be a professional. According to leading figures in the Civil War generation, few professionals in the United States actually met the qualification: another job for university reform.[80]

Even while ideas of this sort were being articulated in the 1870s and 1880s, however, the conceptual framework was changing, without the change being observed very clearly. Some of the same people who had viewed, and who in many ways continued to view, professionalism from the perspective that was just described began to refer to an idea of professionalism that more heavily emphasized career commitments and specialization as contexts for professional behavior, and also to articulate standards of professional behavior more clearly in terms of science. This shift revealed itself in emphases and nuances, in sliding meanings of words and in inferences more than it did in categorical differences, which was one of the reasons that it was difficult to recognize. The education of professionals was increasingly discussed in terms of specific knowledge and skill in subject fields defined as a person's lifework, while other activities toward which "professional" attitudes were appropriate tended to be portrayed as a separate category, to which terms such as "amateur" and "avocation" were applied. For most of the Civil War generation who wrote about university restructuring in the context of broader national reform, this did not mean giving up, for example, liberal arts education as essential for professional education, but rather of making a sharper differentiation between two kinds of education and viewing the first as preparatory for the second. As we shall see in the next chapter, this shift, together with a number of other changes, had a profound effect on the way in which the relationship between the core public sphere and higher education was understood.

Another factor of importance was the construction in the late nineteenth century of two specific professions oriented—one completely and the other in part—toward the public sphere: journalism and university teaching. In the United States until just before the Civil War, journalism had been an occupation but not a profession. Unlike Western Europe, where journalism was at least a well-established path to jobs of more respectability and political significance, America's newspaper journalism was to a large extent a sideline: a temporary expedient or a permanent dead end for literate people in need of a job. Around the time of the Civil War, however, as the opportunities for journalistic employment expanded with the explosion in publishing and as the occupation attracted people of ability and sometimes of significant social standing, attitudes began to change.[81] Newspaper reporters and lead writers as well as editors began to claim special status for themselves because of the responsibility they bore for informing the public and guiding its consideration of issues. The emergence of the reformist or "crusading" newspaper triggered an expansion of claims that the "Fourth Estate" played an essential role in overcoming narrow, interest-governed opinion and exposing the corruption of the public sector by the private. In the process, journalists who made such claims for their roles also developed what amounted to a code of ethics for what they increasingly described as their profession—a career that could be respectably followed in itself, without its having to lead anywhere else.[82] After the turn of the century, the claims of the profession were reinforced by the establishment of schools of journalism at American universities and by the creation of the Pulitzer prizes.

Journalism never became so professionalized that schools of journalism monopolized access to career tracks. The earlier notion of a more informal professionalism maintained itself, especially at the prestigious organs of the core public sphere, defining a career track that led, for example, from the *Harvard Crimson* to the *New York Times* or the *Atlantic* rather than through journalism school. The same was not exactly true of the other new profession, college and university teaching. It, too, had been an occupation but not a profession before the 1850s. Its professionalization took place more or less simultaneously with the reconstruction of American higher education after the Civil War. The profession was built around the adoption, mainly from Germany, of the system of postgraduate degrees that is still the portal of entrance into university teaching. Many decades were required, extending into the second half of the twentieth century, before the Ph.D. became the essential teaching degree in arts and science fields at all major universities, but the direction in which things were going was evident by the 1890s. As we have seen, and as we will discuss shortly, the professionalization of university teaching included the formation of what was called in earlier chapters the "academic ideology"—initially also derived predominantly from German sources. In the nineteenth century, however, the academic ideology incorporated very strong conceptions about professional duties to the public and the importance of engaging in public discourse.

One of the major differences between the academic profession and most of the other professions that reshaped themselves in the late nineteenth and the early twentieth centuries (including medicine and law) was that academia retained a large part of the earlier notion of professionalism: not just the connection to public discourse (which in theory it gradually lost, although not in practice), but also the tendency to accord respect to individuals on the basis of excellence across a range of behaviors and accomplishments rather than conceiving of the task of the professional as a narrow one. The academic profession, almost uniquely, built into its structure and practices a considerable amount of leisure—that is to say, time available to be devoted to tasks other than what was specifically required for teaching, not time to be wasted on inactivity. But whereas, from the 1870s to the 1890s, there was a strong sense that active engagement in public discourse was at least as acceptable as any other way for an academic to use this unscheduled time, thereafter the dominant view held that time should be devoted primarily to research in an academic's specific field.[83] Participation in the discussions of the core public sphere was acceptable, but for the participation to count in any way as a *professional* activity, it was supposed to be the result of disciplinary research, meeting the criteria of being objective and scientific and conducted "for its own sake." We will see how this change occurred when we turn in the next chapter to the relationship between the building of the core public sphere and the reconstruction of American higher education after the Civil War.

The Classic Core Public Sphere

By the beginning of the twentieth century, the core public sphere in the United States had acquired the status, the structure, and the cultural character it has retained down to the present. It possessed a cadre of trained professionals, with attendant ideologies that defined ideal behavior with regard to public discussion. It also maintained connections with the more traditional professions, especially ones such as law, whose members built part of their collective identities around their relationship to public culture and the public conversation. The core public sphere had a clearly recognized and generally quite efficient organizational structure, largely resting on the quality magazines and the networks of contributors, editors, and publishers in which the magazines were embedded. Equally important, it possessed a network of readers—not enough readers, perhaps, to keep the more pointed political journals such as the *Nation* going without some form of subvention, but still a readership of respectable size, many of whom corresponded actively with the journals. The central periodicals of the public sphere did not enjoy the high level of interaction with politicians and the civil service that some of its builders had wished, but connections certainly existed. The core public sphere had become a framework for the articulation of a rich American literary culture, and it had done so much more quickly than most observers at midcentury could have predicted. Associations of extremely varied type had also multiplied, organizing and deepening public culture and public discourse at the national level.[84] All in all, the founders of the classic American core public sphere had succeeded remarkably well. They knew it, and foreign observers acknowledged it. In terms of the extent and level of public discussion, in terms of intellectual sophistication, and especially in terms of effect on social action, what they had created constituted a major change in the collective life of the United States, perhaps the most important aspect of modernity in America after the industrial economy and a powerful nation-state. One of the most impressive monuments to its construction and its success was the Progressive movement, which was first and foremost a creature of the core public sphere.

By the first decade of the twentieth century, the core public sphere had also demonstrated some of its shortcomings. It had divided over the issue of American imperialism, without significant impact on policy, as it would do often in the twentieth century. It had failed to produce a comprehensive, self-conscious description of itself or an ideology that justified its existence and that staked out a significant, legitimate place in American culture and society. Until Habermas, no one in Europe did this either, but at least in Europe one of the traditional understandings of "the public" encompassed essentially the narrow, constructed public sphere of educated people. This was not really true in the United States.

The core public sphere also failed to incorporate the full range of important political discourses of the time. The line between "Progressivism" and "Populism" comes close to reflecting the boundary between what was and what was not acceptable in the public sphere before about 1910. Anything farther to the left than Populism—socialism, for instance—was even less at home in the core public sphere. The same held for the labor movement. Journals published articles *about* these phenomena, not always unsympathetically, and occasionally representatives of leftist tendencies were permitted to contribute—usually as part of "fair and balanced" journal issues that contained opposing views—but radical perspectives were not effectively incorporated into the core public sphere before the first decade of the twentieth century—if then.[85] The same was true of minority perspectives. By the end of the nineteenth century, an extremely able group of educated African American intellectuals and writers had appeared at the edge of the national scene, among whom the acknowledged leader was W. E. B. DuBois. Although many members of the group (especially DuBois) consciously directed their career strategies toward penetrating the core public sphere, they were essentially shut out and forced to create a separate African American version of the same thing—only without the impact. DuBois published a few articles in the quality magazines and a major book (*The Souls of Black Folk*) that was extensively reviewed in them, but even he, one of the most accomplished public intellectuals of his day, was unable to find regular outlets for his work or employment in the institutions of the core public sphere.[86] Educated white women did not experience the same barriers if they came from the right social backgrounds, but the range of those backgrounds was rather narrow until at least after the turn of the century.

In general, probably the greatest limitation on the effectiveness of the core public sphere at the turn of the century was its elitist character. Few of its founders even thought about the question of representativeness—that was something for the democratic political system to worry about. But a meaningful public conversation in a democracy must necessarily include both a wide spectrum of viewpoints and a representative spread of people belonging to the categories into which society is seen to be divided; otherwise, the conversation will lack significance. The range of active participants in the core public sphere was not minutely narrow, but it effectively left out a large part of the population. The trouble that many quality periodicals experienced in staying solvent testified to the limited size of the core's participant reading base. For the core public sphere to exercise the influence its founders sought for it on a democratic nation, it needed to expand its scope: to *democratize* itself. In the twentieth century, it did so—admittedly incompletely and in many respects inadequately. To understand how this happened, we need to turn to an element of the core public sphere that has been noted in this chapter only in passing: higher education.

CHAPTER 4

The Public Sphere and the Construction of the Modern American University

Just as the structures and the (then-unnamed) concept of the core public sphere came to the United States from Europe via the Atlantic public sphere in the nineteenth century, so also did the models for changing higher education in America. With the models also arrived important ideas about how universities and the people associated with them should engage in and support public discourse. We must, therefore, start with these European models and ideas.

Higher Education in Liberal Conceptions of Public Discourse and the State

The conventional story of the European origins of university reform in the United States starts with the revolution of higher education in the German state of Prussia at the beginning of the nineteenth century. It was the University of Berlin, established in 1809 and the center of that revolution, that supplied the principal organizational and curricular models for America and much of the rest of the world. For our purposes, though, the account must begin a little earlier in Prussia, with a famous articulation of the reasons that public discourse is vitally important, of the nature of its relationship to the state, and of the identity and the qualifications of the people who should take part in it. It was not the only such articulation, but it was probably the most comprehensive and certainly the best known. This was the philosopher Immanuel Kant's 1784 essay "What is Enlightenment?"[1]

In the essay, Kant describes "enlightenment" or "enlightening" as the process by which humankind learns to use its capacity for reason by throwing off the chains that have kept it in tutelage—chains that are forged mainly by people's laziness and willingness to let others, usually members of self-interested institutions, do their thinking for them. Because it is difficult for individuals to free themselves from superstition and inadequate thinking by themselves, the process must necessarily be a public one, in which people with ideas are able and encouraged to express them openly in print and conversation and in which others are equally encouraged to criticize them. Only in this way can errors be detected and correct reasoning identified and acted on. Nothing in the process threatens public order or the state. On the contrary, the state should welcome intellectual interchange because, if conducted among the very best minds, it promotes the kinds of governmental, social, technical, and moral improvement that it is the state's chief task to encourage. While the process of developing, airing, and criticizing ideas is going on, strict obedience to the law and appropriate subordination to legitimate authority must be maintained. Actions taken as a result of public discourse must be those of the state itself, or at least must be sanctioned by law. Kant (a Prussian state official in his position as a university professor) was, needless to say, not a revolutionary, but rather an advocate of progress through freedom, and thus a liberal, as the term would later come to be used.

But who are the people who discuss all these important matters? Following standard Enlightenment practice, Kant frames his argument in universal terms that suggest he means that anyone who is capable of formulating an idea should participate in public discourse. The reader is soon disabused of this notion, however. With a few patronizing remarks, he effectively bars women from joining the discussion. He makes it clear that, even among men, it is the "scholars" *(Gelehrter)* who may be trusted—and *should be expected*—to engage in public discourse. By "scholars," Kant means men who have graduated from classical secondary schools and have attended universities—the pool from which most of the Prussian civil service was recruited. It is a highly restricted group. Kant gives the example of clergymen of state-supported churches, who are bound to preach to their congregations according to the established orthodoxy of their denominations even as they freely exchange among themselves their criticisms of that orthodoxy. The expected result of the exchange, should consensus appear among the clergy and should the leaders of the state be persuaded of the truth of the consensus, would be a revision of orthodoxy and therefore of the specifications of what is to be preached. But the common members of the congregations are not participants in the discussion; they are the recipients of its conclusion. To involve the uneducated in the debate itself would be to open the way to disorder, to confusion among those not capable of understanding the process. Kant gives other examples of the same sort: civil servants who contribute to a debate on policy in the pages of a magazine, an officer who publishes a critique of army organization in a public affairs journal—in other words, in media read by people like themselves, not the common people. In the real world, Kant implies, the range of participation in the discourse must be limited to those trained for the task. Ideally, there should be a high level of overlap between the trained participants and the personnel of the state—however the state might be organized. The public sphere should revolve around an educated elite, of which a significant part should be made up of civil servants and others with close connections to the government.

This and some other aspects of Kant's essay do not correspond closely to our contemporary notions of liberalism, which have become suffused with democratic ideas alien to Kant (and to most European liberals before the late nineteenth century). They did, however, speak directly to the concerns of people who tried in the nineteenth century to build a modern liberal society in countries with traditions of political absolutism or bureaucratic rule. European liberals generally regarded a public sphere involving members of the civil service as a useful complement to a public that focused on the actions of elected legislatures, and frequently as an essential aspect of effective governance and reform in a rapidly changing world featuring industrialization, urbanization, and class conflict. In this last context, the idea had considerable appeal in the United States after the Civil War.

The kind of individual Kant had in mind, a "scholar" acting as an upper-level state employee or state-certified professional and simultaneously serving as a core participant in the public sphere, qualified for his status primarily on the basis of education. It had been assumed throughout Europe in the eighteenth century that the people who engaged in public discourse would not only be, for the most part, gentlemen, but also that, as gentlemen, they would possess appropriate educations. The knowledge, the rhetorical skills, and the traits of character needed for the civil interchange of the eighteenth-century public sphere were supposed to be developed by the standard classical education of the gentleman, which might or might not be completed at a university. By the early nineteenth century, however, the new circumstances of the revolutionary era, the intellectual and knowledge requirements of the modern public sphere, and the needs of the modern state had all converged to put a premium specifically on university education and to demand its reconstruction. The principal site of the first stages of this reconstruction was Kant's native state of Prussia.

Prussian University Reform and the Public Sphere

The Prussian state reorganized its systems of higher and secondary education and opened the model University of Berlin in a context of deep crisis: Prussia's devastating defeat by Napoleon in 1806 and its subsequent occupation by the French. With the government, senior bureaucracy, and high command thoroughly discredited, Prussian king Frederick William III called to office a reformist, more or less liberal government headed by Karl vom Stein.[2] The new government initiated several major attempts at reform: restructuring the civil service, the government departments, and the army; establishing a system of local self-government; abolishing serfdom; modernizing property law; and granting full civil liberties to all subjects, including Jews. Implicit in most of the reforms was the desire to construct an orderly state that would guarantee individual freedom. Although they considered ideas of full-fledged representative government, the reformers were at least as interested in expanding the basis for public participation in the *discussion* of state decisions. In essence, this meant strengthening and expanding the scope of the public sphere, which they envisioned along the lines that Kant had laid out. They imagined a state that was still essentially bureaucratic at its center but responsive to the public because civil servants would number among the principal participants in open discussion with others very much like themselves: an intellectual elite sharing a common outlook and education. The rationality and effectiveness of the state would be enhanced by the fact that its policies would be adopted after serious, thorough discussion by intelligent and well-educated people conducted openly before a larger literate public. Threats to individual freedom would be averted partly through balancing institutions against each other and through legal protections, but also by the omnipresence of publicity and by the moral and intellectual character of the governmental elite.

The reformers were able to enact only part of their program before Stein was hounded out of office in 1809 and the government became more conservative, but it was enough to change the nature of state and society in Prussia for the rest of the century. Probably the most successful, influential, and long-lasting changes (apart from military ones) were in education. Almost everyone, reformer or not, agreed that effective universities were necessary for an efficient state because Prussia (like most German states) was run by bureaucrats, and bureaucrats were trained at universities.[3] Because the reformers did not want to eliminate the bureaucracy but rather to modify it and the outlooks of its personnel, they had to incorporate universities into their program. And because their vision of an expanded public sphere depended on educationally qualified participants drawn heavily from the civil service and from university-educated professionals, they had to pay attention to the relationship between what students learned in the universities and what was needed to take part in public discourse. These factors created the framework for the changes that were enacted in higher education, with comparatively little conservative opposition.

Many of the changes had already been on the drawing board for some time. Nevertheless, the catalyst for actual change was the appointment in 1808 of Wilhelm von Humboldt as the state official in charge of educational affairs.[4] Despite the fact that Humboldt remained in office for only a little over a year, he managed to push through both the new university for Berlin and a restructuring of secondary schools, to exert his own influence very strongly on the form of the resulting institutions, and to articulate a philosophical approach to higher education that, for better or worse, has managed to retain its impact for two centuries. The university he helped to found today bears his name, with good reason.

Humboldt was one of the most significant intellectual figures of the early nineteenth century, a philosopher and classicist and one of the founders of comparative linguistics. Early in his career, he wrote seminal essays defining liberal notions of the freedom of individuals as against the state and advancing the concept of individual self-fulfillment as the chief goal of society.[5] He was also a career civil servant (and thus an example of one of Kant's "scholars"). Although he had no previous connection with educational thinking or administration, he was able to take the general framework of reform and (with the help of several others, including the philosopher Johann Gottlieb Fichte) give it a particular orientation with regard to higher education. The system that resulted represented a consensus among liberal reformers and progressive education officials.

The basis of the system was the *Gymnasium,* a type of very selective secondary school, entrance to which was to be available only to boys able to pass stiff examinations.[6] That the *Gymnasium* pupils would, in the nature of things, come mainly from the aristocracy, the mercantile classes, and the sons of officials did not bother Humboldt. Means would be made available to locate able boys from other backgrounds and prepare them for the examinations. The key, though, was to exclude students without adequate intellectual capacities (regardless of class) and to use the *Gymnasium* and the university to mold the intellects of the students who were admitted so that, as adults, they would be able to rise above their class origins and any other possible sources of partiality in the pursuit of truth. In the *Gymnasium* they would receive a solid basis in the classical languages, mathematics, and the other subjects they would need for higher studies. They would also learn habits of hard intellectual work. They would then take an examination that, if they passed it, would allow them to attend a university.

According to Humboldt, a properly constructed university would build on the work done by the *Gymnasium* in order to educate future political leaders and civil servants; the larger group of professionals, gentlemen, and private intellectuals with whom educated officials would regularly converse and exchange ideas; and the leaders of culture. They would all be enabled by their educations to examine issues impartially, to apply the appropriate logical approaches and historical parallels to the solution of governmental problems, and to understand the deeper significance of ephemeral matters. They would be trained to be able to do these things *in public,* not just *for the public,* as participants in debate as both writers and readers. Their educations would encourage them to avoid factionalism and party, the most serious pitfalls of public politics.

The university curriculum, modeled by that of the University of Berlin when it opened in 1809, was to be organized to accomplish these ends. Students would make their own decisions about what to study, within quite broad limits. According to Humboldt, having to make choices in the university built a student's intellect and character. After introductory lectures in philosophy and related fields, students were to select a main area of concentration, which they would pursue in depth. It did not matter which subject a student chose, as long as he had an interest in the field and a capacity for it. His choice was limited, however, to a subject that could be considered an aspect of a unitary philosophical enterprise, rigorously constructed but open-ended in aim. Even professional fields such as law, theology and medicine were to be treated in the same essentially philosophical way.[7]

Most important, the curricula in all academic areas were to emphasize original research. The intention behind this provision is often misunderstood. Neither Humboldt nor the other founders of Berlin University believed that student research—to be embodied in a dissertation submitted for the only formal degree, the doctorate—would contribute much to the general store of human knowledge.[8] That would be left to the faculty, to researchers in institutions outside the university, and to more mature "scholars" in the professions and the civil service of the type Kant wrote about. The dissertation was to be an exercise in thinking for oneself, in identifying an academic problem, investigating it rigorously, and coming up with a solution. In Humboldt's view, working at the boundaries of knowledge, even if in a very limited way, developed the intellectual and moral qualities that university education was supposed to promote: among others, those needed by the leaders of the state and society and by participants in public discussion. University faculty were supposed to be especially active in research so that they could both serve as models for their students and make particularly profound contributions to public discussion. All scholarly research, regardless of field, was at least potentially significant for the discourses of the public sphere.

This, at any rate, was Humboldt's theory, or rather part of it—the part relevant to the needs of the public sphere and the state, the needs that constituted the framework within which the university was supposed to operate. There was more, however. To Humboldt, individual research was not just an aspect of preparing people to take part in public discourse and public affairs; it was also an implication of his general neo-humanist philosophy.[9] In his view, the primary life-goal of an individual should be to realize his or her potential to the fullest, to become a complete person, which includes constructing as deep and accurate an understanding of the world as possible. The state should actively encourage the achievement of this goal by, among other things, recognizing the intellect's need for freedom. People cannot achieve self-realization and true understanding of the world by passively learning what others profess. They must create understanding for themselves, which in turn, as a kind of by-product, leads to the advancement of human understanding in general. This was why, according to Humboldt, education should be based at its highest levels on research, on an individual's creation of an understanding sufficiently clear and rational that it can be communicated to others. To Humboldt, public discourse is only partly a vehicle for affording good government and progress. It is also the social activity in which individual self-fulfillment occurs. To a considerable extent, individual development is judged by the individual himself or herself, not by some external scale such as social utility.

This view of intellectual life entails the corollary that, useful as some knowledge may be for public activity or private pursuits, the most significant knowledge is that which is sought for its own sake—in the sense that the search is more important than the result, because it is the search that leads to individual self-development. This did not preclude useful knowledge from consideration in university studies if it could be approached in an appropriate way, but utility was not supposed to have primary influence on the construction of the curriculum. It was far more important that someone destined to take part in government and in public discourse possess the qualities of mind and spirit developed through pursuing "pure" inquiries of essentially any kind than that he take from his university experience a specific knowledge of an "applied" field. Specific knowledge would be obtained later if it were needed, for example in a professional or bureaucratic apprenticeship.

These last points are the parts of Humboldt's theory that were most familiar to educational writers in America in the late nineteenth and the twentieth centuries, the elements that were incorporated into the "academic ideology." The framework within which they were articulated—that of a university supporting a liberal, bureaucratic government, competent professions, and an effective public sphere—although understood well enough at the time of American university reform after the Civil War,[10] tended later to be forgotten. We will see later why this happened. Although Humboldt did not invent the distinction between "pure" and "applied" knowledge (which had largely been alien to the thinking of the Enlightenment), his ideas contributed heavily to building it. Unlike the general scheme for organizing secondary and higher education, Humboldt's notions of individual self-fulfillment and learning for its own sake were not accepted by everyone associated with education reform in Prussia in the early nineteenth century, nor were they later universally accepted in the new style of university of which the University of Berlin became the model. They were, however, prominently discussed everywhere.

Many of the curricular and instructional devices that became standard in university education (the research seminar, for example, and to some extent the large lecture course focusing on the instructor's research) were pioneered in Berlin and subsequently adopted elsewhere in Germany. There were, however, limits to the extent to which German universities (including Berlin) could practically follow the "Berlin model." The focus on research could not be maintained at every level because of cost and the burgeoning size of student bodies, and also because most students were uninterested in research and wanted only to prepare to take the state examinations for the civil service or a profession. And although the emphasis on research greatly changed the nature of advanced studies in law and medicine, it did not become dominant in either field. Many of the researchers who reconstructed science in the 1840s and 1850s in Germany (especially at Berlin University) adopted a radical materialist perspective and rejected some of the basic assumptions of Humboldt's humanism.[11] Together with the redefinition of the academic career around research and publication, this helped lead to a partial restructuring of the concept of university research itself. Advocates of the centrality of research shifted from emphasizing its role in fostering self-development to representing research, whether for the dissertation or the "second dissertation" (*Habilitationsschrift*) required for a professorship, as contributions to the aggregation of knowledge. Employing scientific methods would ensure the "truth" of the knowledge and certify the authority of the researcher. Academic researchers (including those in humanistic fields) presented their work as projects undertaken selflessly in pursuit of the common good of society, the nation, and humanity, not in the essentially individualistic terms that Humboldt had used. Humboldt's ideas about universities and higher education nevertheless continued to be cited. They were used to create an ideology that the humanistic disciplines and the "pure" sciences employed in their (ultimately losing) battle to retain hegemony in higher education in Germany and to keep "applied" fields like engineering from obtaining parity in status and funding.[12]

The actual development of German universities and academia in the nineteenth century occurred, therefore, in changing circumstances that caused them to deviate in some important respects from the kind of higher education projected by Humboldt—even though many of their salient features derived from innovations pushed by him during the era of reform in Prussia. His ideas and those of other theorists such as Fichte remained current, although the meanings of certain key terms ("research," for instance) showed a considerable tendency to shift over time. This was not unlike what happened in the United States in the twentieth century, when a complex set of circumstances caused the realities of American higher education to deviate sharply from the conceptual apparatus that continued to be used to understand and defend it.

One aspect of the thinking that lay behind the Prussian educational reforms remained relatively intact through almost the end of the nineteenth century, although far from uncontested: the relationship of the university to the public sphere. The belief that a university education aimed at producing a fully developed individual remained connected to the notion that it prepared people for public discourse. This belief was extremely prominent among Americans who referred to German examples in the course of discussing and implementing what they called "university reform" after the Civil War. At least as important, but more controversial, was the role of university faculty and students as participants in the public sphere.

It was expected and customary from early in the nineteenth century for German faculty to express themselves on political issues, and they did so. Fichte, for example, was one of the principal proponents of German nationalism during the War of Liberation against Napoleon in 1813–14. A number of problems arose with the practice, however. After 1815, most German governments (including Prussia) adopted conservative, often reactionary, political postures. A substantial proportion of professors were liberals, and a few of them verged on radicalism in the 1830s and 1840s. Their political writings could not help but be critical of the policies of the states that employed them. The same thing was true of some other senior civil servants, who were also, according to the kind of thinking exemplified by Kant's essay, supposed to participate fully in the public sphere. All this resulted in considerable tension during the first half of the nineteenth century, with some dismissals of officials and professors—especially during periods of real or threatened revolution in the early 1820s, in the 1830s, and again after 1848. These led in turn to liberal demands, enacted into law later in the century, that senior civil servants, including professors, be given *tenure* in their positions so that they could not be punished for what they said or wrote by being dismissed or demoted. There were limits and the system did not always work, but it operated sufficiently well that the American reformers of the Civil War era wanted to imitate it in the United States—initially for federal civil servants, not for professors. Participating in the public sphere in Germany also encompassed activities within political parties. This, too, created problems when civil servants (again including university professors) were elected to legislatures as members of opposition parties. In this respect, tenure also proved useful.

As we have seen, it was decided in the United States to go in a different direction when the federal civil service was created in the late nineteenth century: to forbid civil servants from holding national elective office, to limit severely their activities in political parties, and to discourage them from participating as fully in public discourse as was expected in Germany. But, in part because they were not civil servants, American university professors had fewer such restrictions and more expectations about engaging in debate and politics. Thus, although the device of tenure was invented in Germany to protect upper-level civil servants *including* professors when they participated in the public sphere, it was initially adopted in the United States for civil servants *not including* professors mainly in order to undermine the spoils system. After the turn of the century, however, it became one of the principal devices for protecting the public discourse of professors.

University Reform in England

To the Civil War generation in America, Germany was by far the most important foreign source of models for modernizing higher education, but it was not the only one. Because of the working of the Atlantic public sphere, ideas about university reform in mid-nineteenth-century Britain (themselves heavily influenced by German examples) were familiar to readers of quality journals. In fact, the term "university reform" was directly imported into the United States from Britain.[13] British models were dominant, as we have seen, in post-Civil War discussions of the ways in which the core public sphere should be structured. With regard to higher education, they were less frequently cited than German ones until the 1890s, when, for a while, they became fashionable—particularly at the Ivy League colleges. At no time did the remarkable development of new English universities alongside the two old, traditional ones attract a great deal of American attention. The vogue was for models drawn from Oxford and Cambridge; it was fed by the process through which the Ivies defined a role for themselves as elite institutions in modern America and producers of what was expected to be a nonaristocratic governing class, and more generally by the fact that they appeared to offer a framework for preserving the outlines of the traditional American college within the larger structure of a university. These factors were important in the structuring of American higher education, but much less so in the long run than the German models and the ways in which the latter were modified in the United States. British examples did, however, also have a bearing on the ways in which Americans in the nineteenth and early twentieth century viewed the relationship among universities, the public sphere, and the idea of efficient governance for a modern society.

The developments in English higher education that attracted the most attention in the United States derived directly from the ways in which, between the 1820s and the 1860s, Britain moved to employ the power of the state to confront the huge array of problems caused by economic and social change. In the wake of an increasing use of parliamentary legislation to deal with these problems, British politicians and political commentators realized that something else was needed: an administrative apparatus comprised of educated, honest, efficient civil servants to enforce the new legislation, to monitor the changes that occurred in society, and to recommend new state initiatives. The British state included a great many officeholders but nothing like a modern civil service, so Britain set out to create one.[14] The process by which it did so was discussed at length in the quality press and was thoroughly understood by American civil service reformers of the Civil War generation. The builders of the British civil service consulted various models. What they were particularly impressed with in the Prussian political administration was its close connection to the universities, which they recommended that Britain replicate.[15] Not only did the Prussian example suggest how an effective civil administration could be selected, trained and organized, but also how these things could be done (so it was thought) without danger to the underlying structure of authority in society. The key was to establish an elite corps for the civil service and to tie entry into it to the curricula and the types of education provided at the public schools (that is, the major private, classical secondary schools, themselves undergoing reform along German lines) and at Oxford and Cambridge.

Accordingly, starting in one department and moving to another from the 1850s down to the First World War, reformers established examinations for entry into the new, elite ("administrative") level of the civil service that were essentially of the same sort as those given at the universities, in the same subjects, with no pretense to requiring contemporary knowledge or direct relevance to the job the examinee would do.[16] A candidate for a position in the Treasury, for example, could take an examination in Greek and Latin classics, but for many years could *not* take an examination in economics because no degree examination was offered in economics at Oxford or Cambridge.

The logic of this arrangement was explained in Humboldt's terms: specific information was best acquired on the job; university training was supposed to prepare clever young men for thinking abstractly, reading intelligently, writing eloquently, and conversing clearly. It was a fairly short step from Humboldt to Plato: a civil service selected in the new way could easily be seen as the modern equivalent of the Guardians in the *Republic*. Indeed, the most prominent nineteenth-century English translator of the *Republic* was Benjamin Jowett, a hyperactive scholar and academic politician who was a leading theorist of the link between university reform and the new civil service. Jowett exercised considerable influence over American commentators on higher education toward the end of the nineteenth century.[17]

Jowett was a classicist and, during the 1840s and 1850s, a major advocate of reforming the old English universities. Although strongly influenced by German examples, he thought that the Germans put too much emphasis on research and on scholarship in general and not enough on building a well-rounded character.[18] He became a fixture of British public life as master of Balliol College, Oxford, which he tried, with considerable success, to reconstruct as the preeminent model of what a college in an English university should be and as a training institution for the governors of the British Empire. The key to a Balliol education, besides the building of character that was supposed to occur in the supervised, masculine society of the college, was the tutorial system. This had always been a basic instructional device at the old English universities, but Jowett transformed the system by raising the qualifications for being a tutor and insisting on a substantial amount of student contact, with close attention to preparing students for honors examinations. Along with Macaulay, Jowett also served on the commission in the 1850s that established the selection criteria and the justification for what became the Indian Civil Service, the exemplar of the kind of system that eventually extended to all government departments. He also contributed heavily to building the system of civil service examinations based on university subjects.

Jowett's aim was to create in the "administrative" career level of the civil service a modern version of Plato's Guardians—an elite of the "best" who would, like Humboldt's university graduates, be trained to stand above interest and identify the truth in any set of circumstances. This was what an intelligent gentleman was supposed to do.[19] Jowett was not interested in developing *expertise* in the gentlemen who would be the leaders of the future. Experts had their places, but they should not normally be allowed to rise to the upper ranks of government departments where policies were formulated for approval by ministers. The ministers themselves, politicians who were members of Parliament and had risen through their professions and their parties, would naturally be a mixed bag, but Jowett presumed that the best of them would be drawn from the same university pool as the upper civil service, educated in the same way, and perhaps most important, possess long-standing personal relationships with the civil servants, relationships first developed in school or university and maintained through club or association: the original "old boy network." This was the way in which the public sphere would fit in as well. All of the people with whom Jowett was concerned would be regular readers of the quality newspapers and magazines, and (ideally) most of them would be contributors of articles, authors of books, or at least writers of letters to the *Times*. The better magazines and newspapers would be edited by people with the same gentlemanly and university background as the administrative-level civil servants and the "better" politicians. Ideas would circulate quickly and opinions would be thoroughly considered by people of high intellectual capacity and proper training. Once a consensus had been developed, it would be disseminated broadly to the public and would inform policy and legislation. In the confusion of interest group politics, incipient democracy, and social conflict, this seemed a formula for order and stability.

Jowett was not the only advocate of educational reform with the object of forming an intellectual elite whose members would dominate politics, administration and the public sphere, but he was the one whose name became attached to the idea. Because the conception was in many respects an idealization of a network of elite relationships in Britain that already existed and that continued to develop well into the twentieth century, it proved to be remarkably effective in shaping thinking about the appropriate way to organize a modern society.[20] It helped to inform one version of the theory and practice of higher education in Britain in the late nineteenth century and throughout a substantial part of the twentieth. This version was, at intervals over the same period, attractive to Americans seeking to achieve similar goals.[21] The basic notion was meritocratic, but ultimately, the complexities of modern society and outspoken criticism of what could justifiably be called the amateurism fostered by the Jowett model led to the formulation of a different model of meritocracy: one in which (supposedly) raw intelligence and trained and tested expertise trumped Jowett's gentlemanly, Platonic ideal.[22] The alternative was based on a *technocratic* notion of the relationship between higher education and governance that was quite repugnant to Jowett. In Britain, the difference between the notions was institutionalized. Jowett tried to develop Balliol as a school for Guardians; Sidney and Beatrice Webb and other Fabians in the late nineteenth century established the London School of Economics as a training ground for technocrats.[23] Neither side fully won the competition in Britain until technocracy gained the upper hand in the later twentieth century; internationally, technocracy became far more influential at an earlier date. It, too, had substantial effects on American higher education, although in the various pushes toward meritocracy in the United States in the twentieth century, the distinction between

Jowett-style meritocracy and Fabian-style technocracy tended to be elided. In any event, in the United States, both approaches had to be fitted into a system that turned out to be very different from any of its European forebears. That it was so different was not, however, a result of the deliberate intentions of most of the members of the Civil War generation in America who turned their attention to higher education after 1865.

University Reform as an Issue of the American Core Public Sphere, 1865–90

A central part of the standard historical narrative of American higher education is that the modern American university was the product of a remarkable process of expansion and reconstruction that took place between the end of the Civil War and the beginning of the twentieth century.[24] The narrative recognizes that this process was embedded in much broader contexts of change. Some of the recent work on the subject has focused on relationships between the growth of universities and the formation of a modern American nation-state. Mark Nemec, for example, employs theories of state building to relate the development of the research university to the establishment of a strong federal government with a bureaucratic core.[25] Others have pursued connections to Progressivism.[26] Some scholars have placed university transformation in intellectual and cultural contexts: Julie Reuben has discussed the shift from moral philosophy to science in making curricula and defining the purpose of higher education,[27] while Thomas Bender has demonstrated connections between the development of urban universities and complicated changes in American intellectual life.[28]

Only a few historians, however, have tried to relate the post-Civil War reshaping of higher education to the construction of the core public sphere that was discussed in the previous chapter. Bender links some of the changes in New York City's universities to the development of New York as the center of American public culture and the principal home of the public intellectual. Nearly half a century ago, in his centennial appreciation of the Morrill Act, Allan Nevins remarked in passing that the revolution in American public higher education after the Civil War was closely connected to a parallel revolution in newspaper, magazine, and book publishing.[29] In this chapter, we will see that the modern American university is intimately connected, in both its origins and some of its permanent features, to the process of building the core public sphere in the second half of the nineteenth century.

One of the major reasons that the upheaval in higher education took place at all was that, starting around midcentury, the organs of what was about to become the new American core public sphere adopted the idea of changing and expanding American colleges as a significant topic of discussion. The idea was by no means new. Attempts to change the traditional American college pattern had been made in earlier decades, usually locally and usually with little success. Fairly radical proposals for change—some of them considerably more radical than what eventually transpired—had been advanced from time to time. Thomas Jefferson laid out a model for a relatively new kind of public university in the 1810s, although his plan was followed at the University of Virginia only in part.[30] It was not, however, until the subject of change in higher education was adopted as a recognized issue in the quality media in the 1850s, and even more intensely in the 1860s and 1870s, that the outlines of what was to come were worked out and broad interest developed in pushing innovation forward. Many of the leading educational reformers presented their principal expositions of what they wanted to do in the quality journals. The conceptual framework for the new form of American university education derived directly from the discussion in the core public sphere. The framework effectively excluded other models of educational change that were quite viable, or at least seriously considered, in the mid-nineteenth century. It also retained a larger legacy from the traditional American college than many of its proponents initially intended. Although the framework was significantly modified around the turn of the century, partly by the expansion of public universities and the influence of Populist ideas, much of it remained intact throughout the twentieth century.

Americans writing in the new quality magazines borrowed the term "university reform" from Britain and began, in the 1860s, to apply it to the idea that American higher education needed to be restructured. This idea was a central part of the project of the organizers of the American core public sphere. It was a "reform" of the same kind as others that they emphasized: municipal restructuring, redefining the powers of the federal government, and building a modern civil service. These causes suited the conventions of responsible public discourse; they appeared to rise above party and passion and to be amenable to frank but civil discussion in the intellectually upscale press. They were not messy and inflammatory like abolition or its successor, Reconstruction. People might spill ink over them, but never blood. They were all founded on the perception that the world had changed, that the state and the institutional elements of civil society were inadequate or corrupt and had to be adjusted to accommodate the modern world, and that science, rational investigation, systematic criticism, and ethical judgment would produce the proper modes of adjustment if exercised in public. As the project of "university reform" began to be realized in practice in the 1870s, the term itself was dropped, but the general idea persisted through the 1890s.

We can trace some of the connections between university reform and the building of the core public sphere through individuals. Many of the university presidents whose names figure prominently in the "heroic" narrative of the creation of modern American higher education—for example, Charles Eliot of Harvard, Daniel Coit Gilman of Johns Hopkins, and Charles Kendall Adams of the University of Wisconsin—contributed frequently to the monthlies and to magazines like the *Nation* both before and after they became presidents. (The same is true of skeptics about reform such as Noah Porter, who became president of Yale.)[31] They published their most comprehensive explanations of what they were doing and their debates with each other in the quality general press; they contributed to the same journals on other issues as well. But personal connection went far beyond the ranks of the presidents.

John Fiske, famous in his own day but now largely forgotten, was one of founders of both the core public sphere and the movement toward university reform. After graduating from Harvard in 1863, Fiske built for himself a highly successful career in the public sphere as a popular author of books and articles on history, philosophy and science. Although never a regular member of Harvard's faculty, he worked there as a librarian and taught as an adjunct instructor several of the new courses that were part of Eliot's effort to renovate the curriculum.[32] Fiske helped to initiate the serious discussion of university reform in 1867 in an *Atlantic* article that surveyed the subject very comprehensively.[33] Not everything that Fiske said was new or revolutionary, nor were all his suggestions followed, nor still did he avoid the common tendency of reformers to ignore potential incompatibilities between ideas that he espoused. Nevertheless, it was a significant contribution, delivered in the context of a national public discussion of higher education and very much within the cultural format of the public sphere, to which considerable attention was paid.[34]

Another example is Henry Adams, whom we have already met as one of the founders of the core public sphere. Adams and Fiske knew and admired each other. They belonged to the small but brilliant society that took shape in Cambridge in the late 1860s and early 1870s, a society that also included William James, C. S. Peirce, Alexander Agassiz, and a number of other people who helped to give the United States for the first time a significant standing in the global intellectual pecking order and laid the groundwork for Harvard's eventual international reputation. Although almost all of these people had some connection to the university, none of them thought of himself as being solely, or in most cases even primarily, an academic—even James, who was for decades Harvard's best-known professor.[35] They all regarded themselves as public intellectuals, and they all participated actively in a wide range of public conversations.

In 1870, Charles Eliot, then in the first flush of his attempt to inject innovation into Harvard, offered Adams the post of assistant professor of medieval history—a subject about which Adams knew next to nothing. Adams had just decided to give up his attempt to become a high-level journalist in Washington. The part of *The Education of Henry Adams* in which Adams recounts the offer, the conversations he had with Eliot about it, and the beginning of his subsequent seven-year academic career is one of the best-known and most amusing in the book—and probably one of the most misleading.[36] (Adams assesses his time on the Harvard faculty as a failure, whereas almost all the other evidence suggests that it was a considerable success, a significant and lasting contribution to modernizing American university education.)[37] In fact, Eliot wanted him to perform two functions simultaneously: to cover a period of European history that fell between the two professors in the history department and to edit the *North American Review*. This gives an idea of the relationship between higher education and higher journalism that people like Eliot and Adams envisioned: the reforming university would in this material way support a significant but notoriously unprofitable organ of the expanding core public sphere, while the university would gain a faculty member who was a full participant in public discourse.

The way in which Henry Adams's brother Charles tried to convince him to take the job is interesting: "He said that Henry had done in Washington all he could possibly do; that his position there wanted solidity; that he was, after all, an adventurer; that a few years in Cambridge would give him additional weight; that his chief function was not to be that of teacher, but that of editing the *North American Review* which was to be coupled with the professorship, and would lead to the daily press. In short, that he needed the university more than the university needed him."[38] In Charles's view, Henry's academic career was to be a stage in Henry's larger career in the public sphere: academia and quality journalism leading to a senior position in the popular press, reflecting the idea of a center-periphery structure to be imposed on public discourse by the new core. Also, the university connection would provide a kind of legitimacy that journalism, even of the highest caliber, did not by itself possess.

It is also significant that, in making his offer to Adams, Eliot should have connected the editorship to a professorship in *history,* a subject of instruction that had only recently been introduced at Harvard (and in which Adams had no greater qualifications than in any other). It was there mainly because it was one of the primary intellectual contexts for understanding politics and a significant mode in which to discuss public affairs. A commentator in 1871 suggested that the fact that history was hardly ever included in the standard American college curriculum indicated how far out of touch that curriculum was with the needs of the modern public.[39] Eliot and Adams both believed that instruction in history should be introduced where it was not present, and updated where it was to make it more useful in understanding the contemporary world.[40]

Adams took the job at Harvard. Apparently there was no requirement that he publish in history or accomplish historical research himself. He did nevertheless become interested in actually writing history, prompted by the program he designed for serious undergraduates and graduate students along the newest German lines, which of course emphasized research. His interest as a writer turned quickly to American history, in part because of its direct relevance to contemporary political discourse. Adams famously returned to medieval history much later with *Mont-Saint-Michel and Chartres* (published in 1904), but only when he had decided that understanding the modern required comprehending the contrast between the contemporary world and the world from which the modern had arisen. Studying the past for its own sake, divorced from its utility in studying the present, was never part of Adams's view of history as an enterprise.

Although much of Adams's short period at Harvard was taken up with editing and with learning medieval history from scratch, he managed to introduce a large number of changes in history instruction that were significant for both Harvard and American university education in general. This was another task Eliot had apparently laid out for him. "I am," Adams wrote an English friend, "responsible only to the college Government, and I am brought in to strengthen the reforming party in the University, so that I am sure of strong backing from above."[41] The changes Adams introduced were not entirely unprecedented. American instructors in science had already begun to do away with the core pedagogical method of the traditional American college: the daily, graded recitation based on a standard reading assignment. Adams adopted the approach he had seen (or at least heard about) as a student in Germany: lectures, with questions to students and discussions of topics. He heavily emphasized German models overall, for both constituting history as a field of inquiry and organizing the modes of instruction.[42] Adams also helped to introduce the seminar to the United States—not because he had experienced one in Germany (he hadn't), but because he had read about seminars and they made sense.[43] In cooperation with the Harvard Law School, he organized a seminar aimed at people studying law who had completed an undergraduate degree and wanted to do historical research. One of his first students in the seminar was Henry Cabot Lodge, who became a close friend and one of the first recipients of a Ph.D. in history from an American university. Like Adams, Lodge went on to make a place for himself in the public sphere, but unlike Adams, he managed successfully to enter politics. With Lodge eventually serving as his assistant, Adams introduced the first course in *American* history at Harvard—pretty much the first university course in the subject anywhere. In the course, Adams deliberately encouraged his students and especially Lodge to express political views and historical interpretations that differed from his own.[44] Adams's teaching reflected his conscious intention to restructure the mode of history instruction and to use it to model ideal behavior in the public sphere.

Adams resigned from Harvard in 1877 and tried to resume a career at the intersection between the public sphere and politics. He continued to write for the *Nation* and other journals, authored two anonymous novels on political themes, and waited for an appointment to office that never came. He became essentially a full-time historian, one of the first full-fledged "scientific" historians of American politics—which, as we have seen, was fully consistent with the public role he had laid out for himself and with the European models for the core public sphere.[45]

There were also structural aspects to the connection between university reform and the core public sphere. These were heavily influenced by European models. One of them, as we have already seen, revolved around civil service reform. Many of the founders of the public sphere expected that a modern federal civil service would not just afford efficient government and avoid corruption. At its top levels, it would consist of an elite establishment composed of highly educated people who would share the making of policy with political appointees and elected officials. Civil servants would not only carry out their official duties but also, in their personal capacities, regularly participate in the discussions of the public sphere. They would write for the quality journals, they would take part in conversations with journalists, academics, and politicians—as envisioned in the Jowett model and also in the ideal type of the Prussian civil service. Their educations would come in large part from universities, but some would be institutionally connected to them as well. People like Godkin and Henry Adams appear to have envisioned a kind of continuous exchange among members of the upper branches of the federal civil service, the faculties of colleges and universities, the editors and writers of the quality journals and newspapers, and educated politicians.[46] This was not, as we have seen, how things turned out. Since the nineteenth century there have been only a few senior civil servants who have managed to maintain something like the connections that the founders of the core public sphere envisioned: George Kennan, for instance. Government departments have made heavy use of the expertise of university faculty and researchers in particular fields, a connection that has sometimes resulted in reports and studies being placed before the public. But this was only a part of the broader participation of civil servants and university faculty that had been expected.

Henry Adams's employment by Harvard illustrates a significant way in which "reformed" universities were expected to provide institutional support for the public sphere: by giving jobs to writers and editors who conducted high-level public discourse. As the specific terms of Adams's employment also indicate, this role was seen as one substantially separate from that of doing research and providing data for public dissemination and discussion—the job that today is most often assumed to be the principal contribution of universities to public discussion. The latter idea was by no means absent in the late nineteenth century. As time went on and universities worked out their relationships to the state and to powerful interests, the image of the "neutral" university staying to one side and providing information seemed more useful to university spokespeople appealing for donations and justifying budgets.[47] But initially in the post–Civil War period, it was widely expected that universities should employ at least some of those who engaged in national discussions, not as experts claiming professional access to scientific truth but rather as people with reasoned opinions writing for publications of general interest rather than purely academic journals.

Of course, there *were* no academic journals to speak of in the United States until the 1880s, except for a few entirely scientific ones. Almost all significant discussion of literary, philosophical, historical, and social matters (including most of the important debates on education), took place in the pages of the general quality magazines such as the *Atlantic,*

Harper's, the *North American Review,* and the *Nation.* In the 1870s and 1880s, the people who were constructing the new universities expected at least some of the faculty they hired to publish in the qualities and monthlies. The *North American Review,* as the Henry Adams instance shows, was indirectly supported by Harvard, and the *Atlantic Monthly* and the *New Englander* were understood to be standard outlets for members of the Harvard and Yale faculties respectively. Even in matters of faculty hiring, the connection between the reforming universities and the new public sphere sometimes manifested itself. For example, President Gilman of Johns Hopkins consulted Godkin in 1882 about the qualifications of the economist Richard T. Ely for a teaching position, believing (incorrectly, as it turned out) that Ely was a regular contributor to the *Nation.*[48]

The period of close association between universities and journals with general readerships did not last long, however. Universities created or restructured along the new lines took an active role in establishing scholarly and scientific journals from the 1870s on and dropped their informal attachments to magazines without professional standing. University faculty comprised most of the editorial boards and most of the contributors to the academic journals. The journals themselves generally belonged to disciplinary associations, as in Europe, rather than to specific universities, but universities supplied financial support through subscriptions and often by direct or indirect subventions (as they still do).

The last point indicates another way in which institutions of higher education were supposed to support public discourse: through their libraries. Thomas Jefferson had suggested much earlier that public libraries be established to make knowledge generally available and that properly qualified nonstudents be allowed to use the library of the University of Virginia, but little was done about either suggestion.[49] For years, college libraries were insignificant. After 1865, however, under the heavy influence of German examples, university reformers moved strongly to expand both the size and the importance of university libraries so that they could support faculty and student research, promote the creation of the range of journals thought to be necessary to encompass the varieties of science and the scope of public discussion, and provide a resource for the educated part of the community. This directly paralleled the revolutionary spread of public libraries. From the late nineteenth century to the present, purchase of subscriptions and books by libraries has been the cornerstone of the publishing business in fields outside the range of profitability in the open literary market.[50]

Following an English rather than a German model, the new-style American universities also began in the late nineteenth century to establish university presses to publish books on scholarly subjects and on subjects of concern for public discussion. This amounted to a direct subsidy for scholarship and for the public sphere, since most university presses were not (and are still not) self-supporting. The importance of university presses for the public sphere is often overlooked. It does not, for the most part, arise from the fact that their output is read by the general public. Their main contribution is to make available detailed, well-researched studies important for understanding complicated subjects, which typically supply a large part of the sources for more popular trade books on their subjects.[51]

Probably at least as important as the direct support universities were supposed to give to the core public sphere was their expected role as the arbiters of *standards* for public discourse and as providers of *legitimacy* for the people who took leading parts in it. This role partly followed European examples, but it took on unique characteristics in the United States. In part, it meant discussing and debating matters with civility—the kind of behavior that Henry Adams attempted to model in his courses at Harvard and that William James and Josiah Royce, who disagreed about a great many things but maintained a strong collegial relationship, famously displayed in the Harvard philosophy department in the late nineteenth century.[52] But it went well beyond that. In assessing his experience as a graduate student at Johns Hopkins in the 1870s, Royce reported that he and the other students had learned to accept searching criticism without regarding it as personal, to see it as part of the process of creating understanding, both individually and collectively. "And I mention the matter here because it suggests one of the most important offices that a University has to fulfill, that of teaching its scholars, *and through them the general public,* how to bear without malice and without rebellion, the plainest of parliamentary speech in matters that concern the truth. *Only the academic life can teach a nation the true freedom of enlightened controversy.*" [Emphases mine.][53] The lesson is not just for scholars in their closed-off world; it is for the entire public. This has been an essential element of the public sphere since the eighteenth century, seldom fully realized but nevertheless significant as a prescription for effective public discourse and actually practiced to a surprising degree.[54] In the United States, universities are supposed to be among the primary locations where the prescription is most faithfully followed and where it is most seriously modeled for students and for the public in general. The legitimacy of public debate is emphasized when it is undertaken according to academic procedures and when it takes place in academic surroundings (hence the tendency to hold major debates during presidential campaigns at colleges and universities).

In the minds of the post-Civil War generation and its immediate successors, one of the principal cultural patterns through which the ideal of frank, civil, rational and impersonal criticism was to be realized was *professionalism*. As we saw in the previous chapter, professionalism meant more than just expertise, although that was important and was certainly one of the things that reformed American universities were intended to provide. Professionalism was also seen as an idealized behavior pattern and state of mind. It incorporated the notion that the members of a true profession possessed a moral status that made them trustworthy, not only for individual consultation but also for the performance of public duties. By the latter part of the nineteenth century, proponents of university education regularly argued that the established professions in the United States could not be regarded as "real" ones, in terms of either the individual qualities of their members or their outward presentation as performers of public duties and contributors to public discussion, until the majority of practitioners possessed liberal educations as well as specific training in their fields.[55]

Curriculum Change, University Organization and the Public Sphere

Many historians of education have written about the changes in organization and curriculum that took place in American colleges and universities in the late nineteenth century, and also about the foundation of new universities, new academic departments, new degrees and new disciplines.[56] There is no need to repeat the narratives of institutional history here. What I want to do is to survey some of the issues having to do with university curriculum and organization as they appeared in the quality media of the last third of the nineteenth century, where, as we have seen, the most significant debates over these matters took place and where many of the ideas that shaped institutional changes were principally articulated. I will focus on aspects of the issues that in some significant way involved the relationship between higher education and the core public sphere.[57]

One of the central issues concerned the model of the traditional American college. Apart from the questionable quality of instruction at many of the existing small colleges (a significant matter in the minds of the reformers, to which we will return), people writing about higher education debated about whether or not the model should be retained at all, whether the college curriculum should be totally discarded or merely modified, and whether or not there was merit in the practice of having students live in the institutions where they were educated, under close supervision. The majority of the advocates of university reform in the United States from the 1860s through the 1880s saw relatively little of value in the standard American college. Most were willing to accept the continued existence of some colleges as a practical matter, since it was unlikely that they could be totally replaced, but on the whole they favored reconstruction along what they saw as radically different "German" lines.

Before the changes of the post-Civil War era, the standard American college curriculum was fixed.[58] During each segment of the four-year course, everyone in a class (freshmen, sophomores, etc.) studied the same things—together. In the normal liberal arts college, students started with classical language courses (supposedly based on extensive previous study of Latin), continued through a smattering of other predetermined subjects and a fair amount of completely abstract mathematics, and finished in the senior year with advanced subjects including moral philosophy, which was typically taught in the smaller colleges by the president. Instructors graded students on daily "recitations" in each subject. Colleges totaled the grades to measure yearly standing; at the end of four years, the faculty put students in numerical class order for graduation. The subjects studied were nearly the same everywhere (in theory at least) for each of the major types of institution. The new "scientific schools," some of them independent but most attached to colleges as at Harvard, Yale and Dartmouth, were more flexible but still adhered to a standard pattern of year-by-year instruction. Critics claimed that American colleges were not institutions of higher education at all, but rather glorified secondary schools.[59] Defenders replied that they were ideal institutions for molding character because of their emphasis on moral education, because of their residential character and close supervision of student learning by the faculty, and because of the sense of corporate belonging that they developed through intramural institutions—especially the four classes into which the student body was divided.[60]

German universities were quite different, in both theory and practice. Students had a very wide range of choice about what they studied and when they studied it. Because it was assumed that they had received thorough training in languages and mathematics and other basics in secondary school, they were expected to concentrate their university studies.

Instruction was carried on in lectures (attendance at which was only theoretically required) and, for advanced students, in research-oriented seminars. The principal learning activity revolved around a student's private reading, which was guided by syllabi and lists of texts to be covered in general examinations but was expected to include a large amount of material chosen by the student. Choice was the key: the freedom to choose was supposed to help develop an autonomous intellect and appropriate character, as well as the qualities needed in a good citizen. To American university reformers such as Eliot and Gilman and a host of contributors to the discussion of higher education in the organs of the core public sphere, this kind of freedom was the essence of the "university" as opposed to the college.[61] It extended to living arrangements. Students at German universities found their own lodgings; universities were not residential institutions. Students were subject to some disciplinary control, but as American observers pointed out, it was generally quite lax.[62] German educators rejected the notion that the characters of persons of university age could be beneficially "molded" by close residential supervision of the sort appropriate in secondary schools.

Even at the level of institutional models, the discussion of higher education involved considerations of the public and of public discourse. In fact, both sides agreed quite explicitly that the principal context for debate was the literate public deciding for itself what higher education's public responsibilities were.[63] An anonymous advocate of reform along German lines writing in the magazine *Galaxy* in 1871 argued that the United States could no longer cite its youth as its excuse for an inferior system of education. "A nation that has been rebaptized in fire and blood and has saddled itself with a heavy debt has already passed its adolescence." Moreover, rapid changes in transportation have placed the United States in immediate proximity to "the ripest culture" of Europe (and, by implication, to the greatest military threats). Americans have as much need as Europeans of the most advanced education because "they have the same vast legislative, social and financial problems to meet and solve."[64] The author thus situates himself firmly in the outlook of the Civil War generation as described in Chapter 3. He defines reform (in this case, of higher education) as a necessary response to the general challenges of modernity and to America's arrival as a powerful nation-state. What is required under such circumstances is leaders capable of the intellectual effort of thinking about contemporary problems, discussing them, and coming to a consensus about solving them. They need to be, above all, "leaders of thought." They should possess two intellectual characteristics that must be provided by an appropriate education in a real university: excellence in one field, and a "general culture" that will allow them to "sympathize with and appreciate [their] co-workers in every other department." He makes it clear that the core of the people he is referring to is made up of professionals: "lawyers, doctors, editors, teachers, writers."[65] Note that the list explicitly includes those who are professionals in the public sphere.

Debates over the form of institutions of higher education merged with debates over issues concerning curriculum. The latter also heavily emphasized the public implications of what was taught in colleges and universities. Noah Porter, later president of Yale, defended the traditional college curriculum on that grounds that it prepared young men for performing "public duty," by which he meant not "merely professional duty, but duty in that relatively commanding position which a thoroughly cultured man is fitted to occupy." A cultured man is one who "has been trained to know himself in his constitution, his duties, and his powers; to know society in its history and its institutions, and to know nature in its development and scientific relations."[66] Fair enough, responded the reformer writing in the *Galaxy*, but where in the traditional curriculum does knowledge of history fit? He says that as of the present (1871), only three American colleges even attempt to teach it (the chief reforming institutions: Harvard, Michigan, and Cornell). He claims that "our schoolboys know more of history than our collegians." And then the crucial point: "The result we read every day in the nauseating ignorance displayed by our editors, lawyers, would-be orators and critics, to say nothing of the small-talk of society."[67] The author views history as the significant body of knowledge, the principal conceptual framework, for public discourse. That was the reason that Eliot and Adams were using the study of history as a wedge to pry open the curriculum at Harvard. The need for an effective public sphere thus constituted a major basis of the argument for overturning the traditional American pattern of college education.

It was not just the absence of fields of knowledge needed for the public conversation that condemned the traditional curriculum from the standpoint of the reformers. They also decried the absence of training in the theoretical foundations of public as well as private endeavor. They argued that the professions, the key groups for such endeavor, required a framework of ideas that gave meaning to the details of practice. They claimed that in the United States, college provided no such framework. It does not, complains the *Galaxy* author, teach principles of law to the future lawyer, political economy, history or aesthetics to the writer, or much of anything to the teacher. German universities, in contrast, although they make no attempt to teach the practical details of professions, nevertheless emphasize investigation of the theoretical principles on which the professions are built. In Germany, future practitioners of one profession are not prevented from attending lectures in other fields, especially general ones useful for participation in public affairs or intellectual life. Students follow their inclinations and their perception of what might be significant for them as members of the discursive public, while they prepare rigorously for qualification in their specific fields. The result: the most efficient of modern states, constructed by a nation that only a few years ago was generally discounted among the powers. "No other nation can produce on an emergency so many men trained for just the work that is demanded of them, be that work the marshalling of a vast army, or the building of a railroad, or the revision of a code of laws, or the publishing of a Sanscrit dictionary, or the illustration of a young ladies' poetical album."[68] The *Galaxy* author is quite literally bringing out the big guns of 1871: Germany's smashing success in the recent war against France.

Perhaps the most heated and best-known issue concerning university reform in the 1870s and 1880s had to do with the extent to which the unitary, lock-step curriculum of American colleges ought to be altered in favor of the variety of options for study available in Germany and the freedom German students enjoyed to choose among them. The extreme view among the reformers who dominated the discussion in the core public sphere was stated by Charles Eliot. Eliot thought that specific subject requirements for the bachelor's degree should be kept to a minimum, that the offerings of a university should be packaged as particular subjects taught in courses (usually in the form of lectures) by professors who were experts in the subjects, and that as much as possible, students should be allowed to choose whatever courses they wanted to take.[69] Eliot did not invent individual courses or optional subjects. They had already been introduced elsewhere, especially by Philip Henry Tappan at the University of Michigan in the 1850s in direct imitation of the Berlin University model.[70] However, because of Eliot's position as president of Harvard and his personal prominence in the public sphere, it was his advocacy of "elective" courses as the basis of undergraduate education that defined the concept. This has been a frequent theme in the history of American higher education. Many crucial innovations have had their origins at state institutions—in the nineteenth century, particularly at the universities of Michigan, Vermont, Virginia, and Wisconsin (and also at Cornell, which was and is peculiar in being both public and private at the same time). But what has typically been remembered is the point at which Harvard and a handful of other, mostly private, universities discussed and adopted them—largely because of their centrality in the core public sphere.[71]

Eliot and his supporters at Harvard and elsewhere did not believe in fixed concentrations or "majors," but rather in allowing students to form their own concentrations or not, as their careers and interests moved them. Eliot thought that it was, as a practical matter, unlikely that the English-derived bachelor's degree could be done away with in America, so the German model could not be followed exactly. But to the extent that the secondary school preparation of students allowed (and one of the reasons for reform was to encourage the schools to become more rigorous), undergraduates should be able to study what they wished. Such freedom would not only allow the college to respond flexibly to the varied capacities, tastes and career intentions of students, but it would also encourage students to develop independent, resourceful minds. In addition, Eliot supported the model of professional programs with bachelor's degrees as prerequisites and the creation of doctorates as postgraduate degrees: significant deviations from German examples, but ones that offered the easiest accommodation with the existing American system.[72] These last two initiatives experienced relatively little opposition and became standard. The unstructured "free electives" curriculum, although adopted with modification at Harvard, did not.

Reformers were of different minds about "free electives." The *Galaxy* author favored a little more structure than Eliot did, but on the other hand was not convinced that colleges and bachelor's degrees had to be maintained at all. Many others could not avoid the curriculum reformer's perpetual vice, which is to think that it is possible to specify everything that a student needs to know. Eventually, well before the end of the century, consensus developed around a model that John Fiske had articulated in his seminal article of 1867: a three-part curriculum framework consisting of electives, general education with options within required areas to promote "breadth," and a concentration in one field. (Fiske had believed that electives should be for honors students, but that was dropped.)[73] This ultimately evolved into the standard American pattern, mainly when it was adopted by the large state universities in the late nineteenth and early twentieth century.

The discussion of electives reveals not only the wide range of views that could be expressed by people who claimed to be influenced by the German university model, but also the tendency to resolve differences by including *all* positions (at least to some degree) within the curriculum frameworks that were actually implemented. This tendency extended to the traditional college system, several elements of which survived formally intact—most notably the arrangement of four sequential undergraduate classes and their use in structuring plans of study. Although there were traces of this arrangement in English and German (and also French) higher education, the system established in the United States was not actually consistent in this regard with any of the Europe examples most often cited. It could be, and was, justified by saying that it suited the "American character" or the peculiar circumstances of American society, although such justifications seem to have been mainly post facto constructions.[74]

The varied perspectives on curriculum displayed in discussions of university reform and fitted together into the standard American curriculum actually reflect very different conceptions of the purposes and nature of higher education, including the "Humboldtian" notion that individuals shape their own characters, mainly through intellectual effort, and the idea embodied in the custodial academic institution that characters and intellects need to be molded according to an imposed pattern because most individuals cannot do the job themselves. The differences between the conceptions have never been resolved, nor is there any pressing reason that they *should* be resolved since neither appears to be overwhelmingly more accurate than the other. The fact of their coexistence, however, led to a continuing tension in thinking about higher education that, by focusing on the question of what kind of person, as an ideal type, a university graduate should *be,* diverted attention from the question of what, in terms of the needs of the public, a graduate should be able to *do.* As time went on, this had a bearing on discussions of the role of higher education in connection with public discourse.

Whichever conception of how to form the ideal graduate a particular reformer held, the image associated with the ideal was a masculine one. "Manly" is the term the *Galaxy* author uses most often to describe his version of the ideal. He expresses the reason for affording students freedom to study what they wish in the following way: "After all, there is but one way to make a person a man, and that is to treat him as a man."[75] He is, however, clearly not opposed to higher education for women, as long as they don't live in dormitories near male students or attend the same classes.[76] Indeed, one of the advantages he sees in adopting a "university" system on something like the German model is that it would make the opportunity for real higher education available to women. The author seems to be unconcerned about the possible contradiction here. He appears to think that women at universities can be regarded as honorary men, that the kind of "manliness" he is writing about, characterized by intellectual strength, self-motivation, and responsible adulthood, is something that women can share. The leaders and shapers of the major women's colleges of the post-Civil War period agreed, for the most part. They attempted to follow the new "university" curricular pattern, if not the new model of student freedom, whether their institutions were actually attached to universities or stood separately.[77]

Few of the best-known university reformers of the post-Civil War period came out explicitly against higher education of "university" quality for women, although few of them thought of women as ideal students either, particularly at what we would call the graduate level.[78] As long as an important aspect of university reform was oriented toward preparing students for activity in the public sphere, there was little obvious reason to exclude women even in theory, since women already played a considerable role in public discourse.[79] However, emphasis shifted toward the turn of the century to presenting graduate and professional education as a process of developing expertise and toward making standards for admission to professional education more "rigorous" according to criteria that were aggressively delineated as masculine. Under those circumstances, educated women were systematically excluded from graduate schools, the better law schools, and, after the Flexner Report of 1910, from medical schools and were shunted into professions specifically thought of as suitable for them (elementary and secondary teaching and nursing, for example). What they were *not* excluded from (despite some discrimination) were the public sphere and the undergraduate level of the public universities, as we shall see in the next chapter.

Faculty Professionalization, Expertise, and Research

In defending the existing order of things in American higher education, Noah Porter argued that there was no need for college instructors to be productive scholars and, in any event, that colleges should not appoint professors who were *primarily* scholars. Such faculty, he said, possessed largely ornamental value. The author of the 1871 *Galaxy* article disagrees. He cites the German universities, where professors are researchers who produce original contributions to scholarly and scientific fields that "shed a luster upon the university." He asks whether distinguished people associated with Harvard such as James Russell Lowell, John Fiske, or William Deane Howells are "merely" ornamental or are central to the university's mission. (It should be noted, however, that the examples he cites are all people who spoke to the public as intellectuals, poets, critics, and writers of fiction, not as scholars.) The question of ornamentality aside, he says that the German professor represents learning itself and thereby encourages students to make the effort to learn rather than simply to memorize. He hopes that America will produce a "real [i.e., German-style] university; not a mere dwelling factory such as we already possess by the score, but a centre of thought and research, an exchange-place of learning, a nursery where the young tree, removed from the forcing atmosphere of the hothouse, may develop itself freely and hardily in the broad sunshine."[80] Like others writing in the early years of university reform in the United States, the author expresses a view of the function of faculty research that emphasizes its utility in the education of students and its role in the conduct of public discourse.

The issue of faculty status, very important to most American university reformers, also involved a connection between research and the public sphere. The author of the *Galaxy* article advocates recognizing university teaching as a highly respectable profession whose practitioners are to be admired for their learning, thought of as symbols of the excellence of their universities and their nation, and listened to for what they have to contribute.[81] One of the reasons for Germany's current success as a nation is, he claims, that people of proven ability produce works in their fields that give depth and authenticity to public discussion, and their expressions of their opinions, even on matters that do not directly concern their areas of concentration, are heard with respect because of their intellectual eminence. A professoriate that is respected by the public is an important element of an effective modern society. Nothing of the sort exists in the United States, says the author, but it should. The author represents a researching faculty as professionals: gentlemen who demonstrate their moral standing in their individual lives and in their adherence to the established ethical standards of their occupations. Like other professionals, they should be educated in the theory underlying their fields, and their expertise should consist mainly in their ability to place the practical tasks of their profession (in this case, teaching, research, and participation in public discussion) in appropriate theoretical frameworks. The emphasis in this conception is on the *public* aspects of professionalism, on accountability and engagement, which transcend the boundaries between specific fields. At the same time, the author reveals, here and there, hints of an alternative view, an image of professional researchers as experts in their disciplines, *separate* from the general discursive community and from professionals in other fields. They seek the "truth" scientifically in their areas, and when they find it, they present it to the public in the expectation that it will be accepted on their authority. The author does not seem to recognize the possibility of an inconsistency between these perspectives. Few others at the time did either.

Over time, though, a change in stress can be perceived. In his well-known 1872 inaugural address as president of the University of California, Daniel Coit Gilman emphasizes the role of the university professor as a contributor to public debate and that of the university as, among other things, an institution that prepares students for public discussion.[82] In an article published in 1896, however, Gilman states that the university "must be a shrine to which the outside world will resort for instruction and guidance upon the problems of the day, scientific, literary, educational, political. It must be a place from which are sent forth important contributions to science—theses, memoirs, books. Here every form of scientific investigation should be promoted." Then, to a practical administrator like Gilman, the key point: "Researches too costly for ordinary purses should be prosecuted at the expense of the general chest."[83] The shift Gilman demonstrates was quite general, if not often remarked at the time. Predominance in the discourse of higher education shifted to an image of the university as lying *outside* the public and of its faculty as experts who, from the neutral ground of science, dispassionately supply the results of their work as certified truth. This image is not necessarily incompatible with that of public engagement; in the nineteenth century (as today), a fair number of university faculty projected both, and in any event it is not always possible to maintain the distinction in practice. The obligation of academic researchers is to observe standard conventions about obtaining, employing and citing evidence and revealing interests and assumptions, not to inhabit a "shrine" from which, god-like (or at least oracularly), they reveal truth to the people outside. Even if they wanted to do the latter, the best they could come up with would be a revelation of consensus among researchers about particular questions—assuming that consensus existed. But that was not the idea that the increasingly hegemonic image of the "detached" researching faculty conveyed.

There were many practical reasons for the shift in discursive emphasis. One is obvious in Gilman's statement: the utility of the "separate-and-authoritative" image of faculty and institution in the search for public financial support for university research. The image was consistent with the general emphasis that appeared within reformist and Progressive thinking about segregating mere politics from the making of policy and treating the latter as much as possible as a matter for experts and professionals. Equally important, the new emphasis helped to vitiate some of the fears of wealthy donors and conservative state legislators that their contributions to universities would get into the hands of "reds" of one persuasion or another—a very real source of conflict within American higher education in the 1890s and in the first decade of the new century. The American core public sphere had never shown a great deal of ideological breadth, so that professors who had expressed views within the range accepted by the quality media—between radical Republicanism and laissez-faire liberalism on the one hand and elitist conservatism on the other—could generally be tolerated. But after the Haymarket riot and the red scares of the 1880s, the fact that some academics went beyond the previous range in their public pronouncements raised the question of whether universities should employ such people at all. It was therefore not surprising that university presidents in particular would build their cases around the notion that faculty did "neutral" research that rose above ideological posturing. This did not get the presidents off the hook when particular instances of faculty expressing "dangerous" views arose (the Ely affair at the University of Wisconsin in the 1890s, the Ross dismissal at Stanford a few years later, and several firings during the First World War), but at least it gave them a position from which they could defend both their universities and a particular form of academic freedom (freedom to do "objective" research; research that supported unacceptable political and economic positions could not be objective).

But there was more to it than that. The growing predominance of the idea of objective university research and of faculty as operating outside the public sphere, while it failed to reflect what a large number of faculty actually did and continued to do (that is, participate actively in the core public sphere), corresponded to very important developments besides the practical concerns of university presidents. For one thing, it was connected with and supported by strong positivistic tendencies in academic thought. A great many, probably most, academics actually believed that it was possible, even when dealing with social and human issues, to attain a single objective truth and that it was the universities' particular duty to find it and present it to the rest of society. People who held this belief did so quite sincerely, but in a profession that was just in the process of defining itself as such and was following the German example of basing the claim to professional standing mainly on research qualifications, it was also an important part of the legitimating apparatus of the academic profession. The professoriate could present itself as the priesthood that inhabited Gilman's "shrine" and could on those grounds demand the respect and the support they wanted.

As a result of these and other factors, the idea of faculty as researchers standing apart from the debates of the political arena and the public sphere, providing objective information from outside, became dominant in the discourse of higher education. It also became a significant element of the "academic ideology" in the United States. The notion of university faculty as legitimately participating in public discourse as a professional activity was occluded—not eliminated, because academics in fact became increasingly important in the core public sphere in the late nineteenth and twentieth centuries, but downgraded and partly hidden in the conscious structure within which the relationship of the university to the public was understood and discussed. Together with the tendency to focus discussion of undergraduate education on producing graduates who conformed to a range of ideal types (good citizen, moral individual, "well-rounded" person, and—increasingly favored at the turn of the century—paragon of "manliness"), this lowered the visibility of the connection between higher education and the public sphere. The tendency to organize graduate and professional education as processes that trained specialists with particular forms of expertise also contributed to the same result.

By the 1890s, the initial movement to reform American higher education in the "modern" spirit and in conjunction with the formation of the core public sphere had run its course. It had been remarkably successful, in that it had made higher education a major subject of national debate and a high-priority object of private and state funding. If it had not created exactly the kind of university most of the reformers had wanted (as we have seen, they disagreed in detail about that anyway) or the kind of relationship among universities, the public sphere, and the state that they had hoped for, it had altered forever the role of higher education in American life and society. The reformers had modified some of their views of universities—for example, by adopting British models as supplements to the German ones favored earlier, and by constructing the explanation of university research in a way that emphasized its separateness from the public sphere. Nevertheless, most of what they had advocated had been adopted in one form or another, and their notions of what teaching and research were all about had become standard—as they still are.

Nevertheless, the kind of university reform we have just been discussing did not by itself produce the distinctive, essentially unique institution that was the twentieth-century American university, especially the public university. Left to their own devices, the reformers would have created an American parallel to the twentieth-century elitist European university. But they were not left to their own devices. They were forced not only to make the modifications just noted, but also to accept in practice the necessity of major compromises with divergent conceptions of what universities should be and what tasks they should perform for American society as a whole. This fact had a considerable impact on the relationship between American higher education and the core public sphere, and the *public* universities were at the center of what happened.

Public Universities and the Consensus Model of American Higher Education

Public universities had existed for a long time by the last third of the nineteenth century. It could be argued that the very oldest American colleges, especially Harvard, had been public institutions since their inception, supported (admittedly irregularly) by state treasuries and featuring a significant state role in their governance.[84] Only toward the middle of the nineteenth century had most of these connections been dropped. By the time of the Civil War, the distinction between private and public institutions had been fairly clearly established along the lines that are now accepted. Dividing up American colleges into public and private categories put a very large number of institutions in the second, together with the great majority of students, and a very small number with relatively few students in the first. The subject of whether or not there should be public universities at all was heavily debated by reformers. Some were against them, but probably the majority of reformers who wrote in the quality magazines about the issue favored public universities in one form or another.[85] For one thing, until Eliot turned Harvard toward reform and Gilman and his associates started Johns Hopkins in the 1870s, the universities that were considered to be at the forefront of change were either public or quasi-public institutions (Michigan, Virginia, and Cornell especially).[86] For another, with the nation expanding rapidly to the west, only a government-sponsored effort could make higher education available to the future elites of the new territories and states. But access figured in at best a secondary way in the discussion.

The main factor in reformist thinking about public universities was *quality*. New or expanded and modernized public universities played a central role in a strategy that represented them as ways of dealing with two stumbling blocks on the road to improvement of higher education in general: the poor quality of secondary schools in most states—far from the level needed to provide the new-style universities with students who could do the necessary work—and the recent proliferation in all parts of the country of small denominational colleges with very low requirements for admission and graduation.[87] While raising the admissions requirements of Harvard and other leading colleges would undoubtedly cause a few schools to heighten the standards of their instruction, it would not have much effect on the country as a whole. But if state universities of sufficient attractiveness and prestige were to set high standards, especially in the West and the South, and even better, if states in those regions would establish comprehensive secondary school systems and put them under the supervision of the state universities, the effects on secondary education might be revolutionary. This strategy was in fact followed in several states, with Michigan as the most frequently noted example. Although it did not work out quite as expected, the strategy did strongly shape the construction of the American public high school as predominantly a college-preparatory institution—regardless of the fact that until fairly late in the twentieth century a large majority of high school graduates did not go on to college. The strategy with regard to low-standard denominational colleges was somewhat different. Establishing state universities of high quality and prestige and low cost was supposed to either drive the inadequate privates out of business or force them to improve. Again, although the most extreme expectations along these lines were not realized, there is evidence that by the turn of the century, the strategy had achieved considerable success in both regards.[88]

What the reformers had not anticipated was what turned out to be the crucial, definitive fact in the history of American higher education: the huge expansion in number and size of public, state-supported universities from the late nineteenth century onward. It was this fact that made American higher education "democratic" in what was until recently a peculiarly American way and that incidentally worked to democratize the core public sphere. It was also one of the reasons that a handful of state universities around the turn of the century constructed a comprehensive model for all of American higher education—not (for the most part) because their faculty and administrators wanted to, but because they had to.[89]

The model created at the universities of Wisconsin, Michigan, California, North Carolina, Iowa, and a few other state universities was effectively a compromise. At its core was the array of changes and concepts that had been introduced by the university reformers of the Civil War generation and modified in the ways we have just discussed. But on this core new factors impressed themselves, mostly starting in the 1890s: the influence of Populism, which adopted a view of democratic higher education that had to be taken seriously in states where Populists exercised some degree of power, and the growth in demand for higher education. The latter required public universities to choose between adopting a policy of deliberately limiting access to a small minority of the population or making a series of adjustments in their modes of educating. Although many would have preferred to limit access, the political realities they faced led them in fact to make the adjustments. The Populist effect on higher education was more direct and immediate, mainly because it was embodied in straightforward programs and demands made by Populist leaders.

"Populist" or generally democratizing impulses in American higher education did not begin with the Populist political movement. In the 1840s and 1850s, several different strains of reformism—subsequently put into the shade by the hegemony of "German-style" university reform—had found voice in different parts of the country. In the West, particularly in Illinois and adjoining states, there was a movement to create nontraditional institutions of higher education for the "industrious" classes that would eliminate the "useless" aspects of the traditional college curriculum. They would focus on subjects of importance to workers, businesspeople, and technicians and assist them in their occupations, in their roles as citizens of a democracy, and in their engagement as participants in public discussion. Some proponents of "industrial universities" explicitly intended to break the monopoly on politics and culture exercised by gentlemen with college educations—particularly lawyers: the "talking" as opposed to the "industrious" classes.[90] This movement had considerable influence on the initial versions of the Morrill Act and on proposals to create a "National University," although by the time the former was enacted in 1862, the proponents of more orthodox educational ideas had taken charge of the drafting of the law. The movement's influence waned in the late 1860s and early 1870s as the social image with which it was associated—a united front of culturally identical skilled workers and capitalists—dissolved and was replaced by more fragmentary and contentious pictures of society. In the East, several attempts were made to extend the idea of open public primary and secondary education into the adult years, one of which led to founding of the Cooper Union in New York City.[91]

It could be argued that "democratic" ideas of higher education go even farther back, to Thomas Jefferson's proposals for the University of

Virginia. Jefferson, however, envisioned the university explicitly as a place to prepare leaders, and public secondary education at least in part as a means of selecting the talented elite who would attend the university.[92] This was not far from the thinking of the later mainstream university reformers influenced by the Berlin model. What distinguished the alternative initiatives of the 1840s and 1850s—the Cooper Union and the "industrial university" concept—was the idea that higher education should be available to everyone of normal ability who wanted it or might find it useful. It should aim at educating not (or not only) an elite, however chosen and constituted, but the body of citizens and working people of all classes.

In the 1880s and 1890s, as elements of the Populist movement organized themselves in the Midwest, Far West, and South and differentiated their position from the "Progressive" reformism with which the builders of the new core public sphere and the new universities were mainly identified, some of their leaders developed a coherent alternative view of the changes that were needed in the major structures of American society. The essential difference was that the Populists believed the problems of America could best be addressed by expanding the scope of participatory democracy, while Progressive reformists tended to view broad democratic participation as something that needed to be hedged against or counterbalanced. Both views were subsequently accepted into the political mainstream. Some of the differences between them were resolved; others were merely deemphasized. But the Populist impact in many areas—including education—was considerably stronger than the standard historiography has suggested.[93]

Some Populists brought older radical ideas about higher education out of the closet, attacking existing universities and colleges as elitist and antidemocratic and demanding instead institutions that focused on the needs of practical people, especially farmers. Much of the impetus behind the expansion of "A and M" colleges and the creation of the state agricultural extension services emanating from them came from the Populists. In the long run, however, the most influential aspect of Populist thinking about higher education was its take on the new style of university. Populists generally accepted the model, but they insisted that it be adopted for all sorts of institutions for which the earlier university reformers, by and large, had thought it inappropriate: agricultural colleges, normal schools, schools of mining, and so forth. Part of this insistence had to do with prestige. Populists argued that ordinary people going through a course of education with a focus on practical, occupational subjects should not, in a democracy, be treated as inferior beings. Their schools should have the same structure as anyone else's and the certification they obtained should have the same value. What reformers sneered at (like many of their successors in the twentieth century) as a threat to standards, Populists regarded as essential to maintaining real democracy and avoiding the perpetuation of unjustified class distinctions.

Populist commentators on higher education were also interested in curriculum. In part, they explicitly framed that subject in the context of public discourse. This had been true even in pre-Populist radical perspectives on higher education. Proposals for "industrial universities" in the 1850s, for example, generally included provisions for instruction in useful liberal arts such as history, languages, and modern literature (not Greek and Latin) that would prepare students from the "industrious classes" to hold their own in public conversation with professionals and others trained in the more traditional colleges. Some of the same language was repeated in the Morrill Act.[94] The Populists of the late nineteenth century extended this kind of argument, influenced to some extent by the actual practices of the state universities. They insisted on the addition of general education segments to undergraduate curricula, including courses that emphasized training in expository writing, in history, government and economics, and in other subjects that explicitly prepared students for engagement in the public conversation and equipped them for analyzing issues discussed in the organs of the public sphere. In a period in which the main currents of discussion among the successors of the university reformers were turning strongly toward meritocracy and the production of expertise and in which general education was increasingly discussed as a means of shaping character and training an elite of leaders, the Populists maintained the importance of education for public discourse.

In response to both Populist pressure and rising demand for access to higher education from citizens regardless of their political orientations, several state universities (often reluctantly and under pressure from legislatures) adopted new practices and structures. In general, Midwestern universities, especially Michigan and Wisconsin, led the way. What emerged was, in most significant respects, the American state university of the first half of the twentieth century. Rather than expecting perfectly trained students right from high school, the universities admitted both those who could demonstrate qualifications by examination and a large number of those who were more marginal. The courses of the first year or two served not only as preparation for "major" courses and the higher-level studies of the last two years, but also as instruments for "weeding out" the students who could not make the grade. The lecture—the principal "German" mode of instruction introduced by the reformers of the Civil War generation—became the standard format for lower-level survey (and weeding) courses. The "course" itself became standardized to a greater extent than it had been in the earlier reform period, and the qualifications for entering higher-level courses were stated in terms of the specific lower-level courses that were needed as prerequisites. The use of courses not only as means of conveying knowledge and guiding reading but also of identifying deficiencies in knowledge and skill encouraged the proliferation of tests and short papers that has been a hallmark of American higher education ever since. The necessity of accommodating the needs of students who transferred from one institution to another reinforced the tendency toward standardizing courses and requirements. The adoption of a standard method of assigning "credits" to courses on the basis of the number of "contact hours" they involve arose from similar motives.[95] It was also discovered that, with large numbers of students coming to universities from all over a state and beyond, it was necessary to provide them affordable and regulated housing—hence the development of the large university dormitory and its culture. Since many students came from families with no exposure to the kind of culture maintained at universities (even state ones), it was necessary to create a student-support apparatus, which developed into the familiar administrative structure headed by a dean of students.[96]

By the second decade of the twentieth century, public universities structured along the lines of the larger Midwestern state institutions had become in most respects the standard for higher education in the United States.[97] Despite fluctuations in funding, their numbers, their student populations, and the sizes of their faculties grew, and continued to grow. As the post-Civil War reformers had expected, competition from the public universities forced smaller private and denominational colleges to change their modes of instruction and raise their standards or go under. As some of the reformers had hoped, many state universities became pacesetters for educational innovation. It was not that most innovations necessarily came from them as opposed to the most prestigious private universities (although a great many more did than is usually recognized). Rather, because of their expanding size and general significance, they were the institutions that determined whether an innovation would be generally adopted or would disappear or survive only as a local curiosity. But few nineteenth-century reformers (apart from Populists and a handful of non-Populist radicals) had predicted another effect of the growth of the public universities and their assumption of the role of standard model: the extent to which they broadened, and thus in a significant way democratized, the core public sphere. This is a central topic of the next chapter.

CHAPTER 5

Public Universities and the Democratization of the Core Public Sphere

We turn now to two connected developments in the first half of the twentieth century that profoundly affected most aspects of public life in the United States: the movement of universities from a peripheral to a central place in the structure of public discourse, and the process by which higher education (especially *public* higher education) worked to democratize the public sphere. First, some statistics:

In 1882, 26 percent of the students enrolled in American colleges and universities attended public institutions. In 1900, the percentage was 38; in 1930, 48; in 1960, 59; and in 1970, after a spectacular increase in overall college enrollment, it stood at 75: three out of four students.[1] Between about 1900 and the mid-1960s, total enrollment in American colleges and universities rose from 237,000 to 5.5 million—a factor of 23 times.[2] During the same period, the total population of the United States increased by a factor of two and a half times. Thus, the proportion of the total population attending college by the mid-1960s was nearly ten times larger than it had been at the beginning of the century. The vast majority of the increase in that proportion was due to public institutions of higher education. Between the 1960s and the 1990s, college and university enrollments increased at an even greater rate than ever before. By 1991, total enrollment was estimated at 14.2 million students, of whom just under 80 percent were in public institutions.[3] In the course of the twentieth century, the United States became a country in which, by any meaningful standard of comparison, a huge proportion of the population participated or had participated in higher education—vastly higher than in any other country in the world.[4] Since that time, the growth of enrollment as against population has slowed, but numbers have increased (although irregularly), especially at the graduate level. The percentage of students in public institutions fell slightly, but only to about 75 percent of the total in the mid-2000s.[5] What has fallen drastically has been state financial support for higher education. What has risen most impressively has been the cost of higher education to individual Americans.

As we saw in the previous chapter, the vast growth of American public higher education was anticipated by almost no one in the nineteenth century. It was driven by a great many factors. Some were narrow in scope, such as the establishment of the consensus model of the American university. This provided a pattern that public institutions that were not originally intended to be of "university" caliber (normal schools, agricultural colleges, technical colleges) could adopt to raise their status and to seek to accommodate students beyond the narrow vocational limits initially assigned to them. Other factors derived from broader national trends: rising incomes; rising aspirations with regard to earnings, consumption, and social standing among large segments of the American population; a growing tendency toward professionalization of occupations; and an increasing willingness of state governments to engage in the building of economic and social infrastructure. These factors contributed to a continually increasing demand for university education—that is, for enrollment in institutions granting degrees and offering the experience of a comprehensive undergraduate program—because that kind of education was associated with respectable social standing, with high incomes and professional status, and with full access to public discourse. In many states, traditional wariness about spending public revenues on higher education when private colleges could perform the same function was matched in the twentieth century (although never wholly replaced) by a propensity on the part of politicians and local boosters to regard the presence of a state college or university as a major way in which the state could contribute to local development. This encouraged a proliferation of public colleges, and even more, an expansion of the number of student places at public institutions, that continued ceaselessly (with alternations in tempo) throughout the twentieth century.[6] Thus, the public sector of American higher education, which had been a secondary, more or less corrective element in most mainstream conceptions of educational change during the post-Civil War period, emerged in the twentieth century as the dominant part of the enterprise, while the enterprise as a whole, public and private together, became a defining feature of American society.

Universities as Centers of Public Discourse

Amidst all this expansion, universities as institutions and academia as a social and cultural construction continued to move relentlessly into a central position in the American public sphere, even if that position was only occasionally recognized. They did so largely because of factors that were already apparent in the late nineteenth century and that became even more significant in the twentieth.

In the first place, universities became the principal repositories and certifiers of expertise and the main producers of science in an era in which science was seen as the key feature of expertise. Universities provided the educational experience required for certification in the burgeoning array of occupations whose members sought to reclassify them as "professions." They also did most of the work of improving the educational standards of the traditional professions of medicine and law as those fields were reformed in the early twentieth century. The process was reciprocal: universities that adapted themselves to the trend toward professionalization became major beneficiaries of the process. One of the reasons for the rapid rise of American higher education to international significance in the twentieth century was the responsibility they assumed for structuring professional education.[7] A large part of the growth in the importance of universities in the United States was due to their connection to the professions. To maintain the connection, universities were required to become the institutional homes of professional experts. As expertise became an increasingly important qualification for engaging actively in public discourse, this almost automatically gave the universities heightened significance in the public sphere.

The model case of professional reconstruction occurred in medicine. The Flexner Report of 1910 condemned with devastating brutality most American medical schools and, by implication, the expertise of most American physicians. It recommended that many medical schools be closed and the rest be restructured along the lines pioneered at Johns Hopkins, which meant emphasizing thorough education in the sciences and even more thorough practical training in the techniques of medicine. The report took the public sphere by storm and produced consensus that drastic action was needed. Despite some whimpers of protest, the standards of the entire profession were overhauled; a large number of small, independent medical schools went out of business.[8] The remainder moved overwhelmingly toward university affiliation, partly because of the cost of the new regime and partly in order to connect with science. They also affiliated with large hospitals in order to provide the practical training now required. (The reforms also effectively excluded all but a handful of women and African Americans from the medical profession.)[9] Moreover, because "science" implied not just a body of knowledge available at universities but also research, medical faculties developed a structure and a mode of operation that created direct links between research and practice and between researchers and practitioners, so that every qualified physician could be regarded in some sense as a participant in ongoing research, if only as a reader of journals and attendee at conventions. What had once been a professional engagement in the public sphere was now regarded as an obligatory concomitant of scientific expertise. Prestige in the profession (although not necessarily income) came to be distributed in proportion to the amount of involvement physicians had in research and the closeness of their connection to leading university medical research centers.

Similar changes occurred in other professions. Law had already begun to become university centered in the late nineteenth century, but in the twentieth, the older system of training and certifying lawyers through clerking for established attorneys was replaced wholesale by education in law schools, most of which (the ones that survived reform) were attached to universities.[10] Engineering, accounting, pharmacy, business management, and scores of other fields went through similar processes. These processes were not usually initiated by the changes in medicine, but were accelerated and often given direction by them.[11] And as universities became the principal centers of professional expertise, the pronouncements of members of their faculties took on extended importance, a development that coincided with a strong tendency throughout the Atlantic world toward conceptually disassembling the notion of a unitary educated public and replacing it with a collection of fragments: the academic and professional disciplines around which universities were organized. This in turn had profound effects on the place of disciplinary associations in public discourse—again, reinforcing the identification of associations and disciplines with universities. The American Historical Association (AHA) can again serve as an example.[12] We saw earlier that the AHA was founded in 1884 to be a general organization of people interested in history and a support for public discourse and effective national policy. It rapidly transformed itself in the first two decades of the twentieth century into a professional society dominated by academics.

Much of the association's work came very quickly to focus on defining the professional ethics and careers of academics. The AHA did, and still does, retain a conscious commitment to the public. Like other associations, it did its bit for the war efforts of 1917–18 and 1941–45.[13] Occasionally, it makes pronouncements in general terms about broad issues such as job discrimination. But most of its concerted efforts to influence public opinion or government policy involve matters of direct concern to professional historians, such as access to archives and their proper administration.

Both state and federal government agencies also increasingly availed themselves of the expertise located at universities. Because much of the factual material around which public discussion takes place arises from research done for the government, this, too, had the effect of enhancing the centrality of universities in the core public sphere and their functions as providers of scientific legitimacy for positions on policy. Providing information useful to the state had been one of the functions envisioned by nineteenth-century reformers for the new "German-style" universities they hoped to create. In the twentieth century, however, the function became increasingly central to the public identities and the practices of the universities that focused particularly on research.[14] The various large foundations created in the early twentieth century—Carnegie, Rockefeller, Ford, and so forth—made similar use of university-based expertise, sometimes in constituting their directorial boards but more frequently as preferred recipients of grants, as grant reviewers, and in general as professionals who carried on the work of the institutions.[15]

In the twentieth century, universities took on even larger roles than they had in the nineteenth as providers of financial and structural support for the discussions of the core public sphere—sometimes intentionally, sometimes not. It is true that the growing professionalization of university teaching ruled out the kind of arrangement that existed in the 1870s, when professors could simultaneously teach courses and edit general journals such as the *Atlantic* and the *North American Review*—for the same salary. But other forms of university support for the nonacademic public sphere continued, and new ones were invented. University-employed academics wrote, as they still write, a significant proportion of the articles for the quality journals and books for the nonfiction commercial market; today they constitute a large part of the pool of talking heads for television public affairs and news shows, although more often than not they appear in these media as experts rather than generalists. Universities provided place and occasion for addresses, debates, panels, conferences, and other public activities that served as focal points for the discussion of issues.

University commencement speeches became favored media for making important policy statements. The honoraria for speeches, workshops, and the like paid by universities today make up a not-insubstantial part of the incomes of many writers, commentators, and former officeholders who engage actively in the conversations of the core public sphere.

Some modes in which higher education supported and has continued to support the core public sphere derived directly from aims of the nineteenth-century university reformers that we have already discussed. The vast expansion in the size and numbers of American university libraries that occurred with the shift of higher education toward research also had the intended effect of creating a sustained market for serious publications of all sorts, not just academic ones but also the kinds of journals and books that carry on public discussion. University libraries shared this function with public libraries, but had the particular responsibility of buying books and subscribing to journals that published the results of academic research, making them available for consultation by participants in public debate.[16] University presses (mostly subsidized by their institutions) produced a large proportion of the books of this sort. They allocated their uncovered costs to purchasers (usually university libraries) through the prices they charged.[17] In these and many other ways, the underwriting of important elements of the core public sphere by higher education in the twentieth century steadily increased the centrality of universities in the structure of public discourse.

Universities also continued to function in the twentieth century as promoters of civility and freedom in public discussion, to a much larger extent than had been anticipated in the nineteenth century. This again brought them closer to the center of the core public sphere. In addition to teaching and modeling civility and offering places where unpopular ideas might be expressed and criticized, universities in the twentieth century were among the foremost *defenders* of free speech and civility in the United States. It is true that they have not always been able or willing to perform this job outside the limits of scholarly inquiry, and sometimes not even within. The way in which teaching and research are organized at universities and the tendency of academia to develop hierarchies of perspective peculiar to itself have sometimes meant that ideas with a low rank are not given the opportunity to be expressed. Like other institutions, universities are subject to considerable pressure to punish faculty who say or do unpopular things or to keep people who advocate positions that are repulsive to many from speaking on campus.[18] Sometimes they succumb to the pressure, which is not always entirely external. But however inadequately they may at times have performed their tasks with regard to free speech and civility (especially in wartime), they have on the whole done at least as good a job as any other element of the public sphere. The nation owes them a debt of gratitude that it has neither fully acknowledged nor come close to paying—although I think the time has come when it should be asked to do so.

The principal means by which university faculty members (or at least some of them) have protected their own freedom to take unpopular positions has been the system of tenure—adopted, as we have seen, from Germany, but well after the initial German influence on American higher education had receded. Tenure has become so deeply embedded in the professorial way of life that the reason it exists is often forgotten, or at least half forgotten, as though tenure were primarily a necessary incentive to good teaching or good research.[19] It seems to me that the only legitimate justification for tenure is to maintain a basis for intellectual freedom in public discourse, but that is a crucial justification. To defend tenure from the attacks that are regularly mounted on it, it is essential that the importance of tenure for the operation of the public sphere and the conduct of the national conversation be emphasized, and not, for example, the value of job security in encouraging university faculty to do creative research in unusual fields.[20]

If we wanted to pursue the matter further, we could explore a number of broader changes in American society since the nineteenth century as explanations for the growing prominence of higher education in the core public sphere. One might be the sharp contraction in the range and number of local associations, clubs, and voluntary educational and cultural programs that were once central features of community life in the United States and among the primary means by which people ordinarily connected with the national public sphere. In some ways, the functions of these organizations were taken over by local institutions of higher education, especially public colleges as these were established throughout the country. Another might be the elitist image of many of the periodical publications of the core public sphere, which made them seem distant and unwelcoming to many Americans. One of the major ways in which entry into the discussions of the core public sphere has been made attractive has been through college courses—again, not so much at leading universities as through local public colleges.

With these possibilities, we arrive at a crucial point in the argument of the book. Most of the explanations that have been given thus far in the chapter for the movement of higher education to the center of the core public sphere are equally applicable to private and public institutions. But the changes just mentioned provide a hint of the particular, essentially unique function of the public colleges and universities in the United States: they were the principal agents for *democratizing* the core public sphere in the twentieth century.

Democratizing the Core Public Sphere

As the preceding chapters have made clear, the core public sphere is not inherently democratic. Historically, the concept of a discursive public is closely connected to liberalism, but in the Atlantic world throughout most of the nineteenth century, liberalism and democracy were not necessarily complementary in the way that they became in the twentieth. As we saw, many of the people who constructed the American core public sphere thought of it as a kind of *corrective* to democracy, very much in keeping with Progressive ideas about governmental reform as a response to democratic excess.[21] The reconstruction of American higher education in the late nineteenth century reflected a similar outlook. The new type of university featured a meritocratic elitism laid atop the gentlemanly elitism of the antebellum American college, in both respects finding legitimacy in European models. It was not that most of the people who thought this way were actively opposed to democracy (although some were).[22] They believed, rather, that democracy required help from talented elites qualified mainly by their educations. Within this context of creating a public sphere and building higher education, genuinely democratizing features figured as add-ons.

Nevertheless, democracy could not be ignored. For one thing, democracy was the strongest constituent of American political culture. Throughout the late nineteenth and the twentieth centuries important democratic movements, starting with the Populists, insisted on being taken seriously at the highest levels of public discourse and on requiring that higher education take democracy equally seriously and in a positive way.[23] For another, the actual structure of public discourse that developed in the late nineteenth and the twentieth centuries did not neatly organize itself into a hierarchy of influence in which the intellectuals and the high-brow press of the core public sphere could always set the keynotes for discussion and politics. And for reasons of income if nothing else, institutions of higher education—even the most prestigious of them—were unable to maintain themselves solely as educators of the intellectual elite.

In modern societies, diversity demands democracy as a practical necessity—a necessity that overwhelms theory and neat schemes of control that attempt to contain it. Corporate businesses and political parties learned very quickly that they had to accept and work with the messiness of broad, socially and culturally varied participation in the life of the national community in order to have a chance of achieving their goals. The would-be managers of the public sphere—publishers of quality magazines and newspapers, university presidents, academics, professionals, intellectuals, and college-educated journalists—learned more slowly and acknowledged much more grudgingly that they had to do the same thing. They tended, and have tended to this day, to express their resentment of this reality by blaming it all on "commercialization," by saying that that a business sector that depended for its success on "mass marketing" was imposing its values on the whole of American culture.[24] There was, and is, considerable truth to this assertion. But a substantial amount of what they complain about arises from the fact that commerce created most of the models and designed most of the media through which large, diverse publics are brought into general frameworks of communication—admittedly frameworks very restricted in scope, constructed with the intention of influencing a limited range of behaviors and attitudes (that is to say, consumption of commodities). These are the media that, for want of alternatives, the personnel and institutions of the core public sphere had to employ in order to function in the actual world of a diverse public. When someone asked Joseph Pulitzer why, if he preferred to read the (then-) intellectually respectable *New York Post*, he didn't try to make his "sensationalist" *New York World* more like it, Pulitzer reportedly replied, "I want to talk to a nation, not to a select committee."[25]

There were exceptions, of which the most important was higher education. Universities and colleges protected, reinforced, and expanded most of the values and functions that were supposed to characterize the core public sphere. For people who thought they had something to say about matters of political, social, and cultural importance, universities gave them a place to say it and, in connection with the other elements of the public sphere, the means to get it to whatever audience was willing to hear it. Universities also gave them an audience immediately at hand in the persons of the students. By legitimately manipulating degree requirements, it was possible to make sure that most students were at least exposed to the main currents of public as well as academic discourse, although the latter—which did not necessarily emphasize immediate relevance to contemporary issues and sometimes denigrated their significance—took precedence. Higher education afforded access to a public of respectable size that did not require employing the means of mass commercial communication.

Universities did not just disseminate ideas and information. The discussions that took place in classes or afterward, whether between instructors and students or among students by themselves, constituted an increasingly important part of the public conversation itself. For some students, university studies were preparation for later engagement, sometimes as producers and analysts of ideas and most frequently as thoughtful readers and occasional discussants. For others, it was the highpoint of their involvement in serious conversation, seldom again to be approached in intensity—whether from lack of interest or because of the scarcity of structures outside the university that continuously supported the public conversation.

The important points about *public* universities in this context are two. First, by virtue of their relatively low cost and their rising share of the growing pool of postsecondary students, they engaged an increasingly large sector of the population in the public sphere. Second, and probably more important, from quite early in the twentieth century they reflected, if not the exact statistical distribution of Americans among class, race, gender, and other groups, then at least the range of such groups. Simply by functioning, they worked to democratize the core public sphere in these two respects. Even before the 1960s, when federal and state governments began to acknowledge building student diversity as a formal objective of higher education, public universities and colleges had become relatively diverse because of their cost and the fact that (with regional variations) they were open to all who qualified, and also because they offered a wide range of educational choices within the dominant organizational pattern of the university. One of the characteristics of public higher education that was most severely attacked by some critics throughout the twentieth century—the inclusion of "vocational" subjects among degree concentrations—helped to make public institutions diverse. People with a wide variety of career intentions and from very different social backgrounds were brought into the conversations that went on in universities and had the opportunity to learn (whether they chose actually to do so or not) the language of the core public sphere.

We can obtain an idea of what this diversity implied if we look at the changing circumstances of women in American higher education over the past century and a half. Public universities did not invent women's higher education, nor were they the first colleges in the United States that men and women could attend together on relatively equal terms.[26] They were, however, the institutions where "coeducation" first became standard and the common experience of most students in American higher education. The disproportion of men to women in public universities declined fairly steadily throughout the twentieth century, until in recent times the number of women students equaled and then surpassed the number of men.[27] In an aggregate sense, if we consider "women" as a category of the population denied full access to citizenship and the public sphere, the public universities clearly contributed to the democratization of both polity and public discourse by making higher education available to women. It was not just that public higher education provided women with expository skills and access to the media. Because the intellectual interactions that took place in and outside college classrooms were a central factor in constituting the public sphere itself, the participation of women in those interactions necessarily had an effect on shaping the public conversation. Several generations of Americans of both sexes entered the public sphere as writers, discussants, and readers in a context that not only included women but in which women were rightful, active participants in the normal practice of discourse. Coeducation was not the only context for this phenomenon. In women's colleges, some of them public but most of them private, students could develop the skills of public discourse without the omnipresent assumption of male hegemony.[28] There were other structures within which similar developments occurred (in political organizations of various kinds, for instance), but not very many and not on as wide a scale as in colleges and universities.

"Women" is obviously a meaningful category for social analysis and for social action, but women do not act in society only as women. A woman's relationship to the polity, to higher education, and to the public sphere is, like a man's, mediated by class, age, income, occupation, religion, location, and a great many other framing categories. Women's experiences, like those of members of any other segment of modern society, are multifaceted and multivalent. This is one of the things that "diversity" means. Diversity in the sense of multiple, intersecting categories that frame experience is a central feature of modernity. It is probably the feature that makes democracy an unavoidable necessity, because broadly participatory institutional structures and practices appear to be the only ones that can reliably facilitate the creation of consensus among the fragmented categories to which people in modern societies belong.[29] But diversity cannot be adequately accommodated by a democracy that recognizes the existence only of very broad categories or of ones that are treated as mutually exclusive or hegemonic. One of the services that a democratic public sphere performs is to register the complications of diversity—not just by giving voices to people representing the broad categories, but also by incorporating in the public conversation a range of people that covers most of the intersections of significant categories. In the specific case of women in relation to the core American public sphere, it is not sufficient to consider just the relationship of "women" to higher education and to the American public sphere since the nineteenth century. We have to recognize the different experiences of women from different class, occupational, ethnic, and racial backgrounds. To make the point, we can focus on the factor of *class* as it intersected with the social category of "women."

The core public sphere was from its beginnings relatively open to contributions from educated, upper-class (or at least well-connected) women. Women such as Louisa May Alcott were among the most prolific and successful novelists of the nineteenth century—a century in which the novel was a major medium for linking the elite elements of the public sphere to a wider audience. Many of the new magazines of the second half of the century published large numbers of articles by women, several of whom became more or less regular contributors to particular journals.[30] Among the places where the mainline, respectable suffragists found an early audience was in the pages of the quality journals, and much of their effort was directed toward making their case to the audience of the core public sphere.[31] Few of these women had college educations; they were educated by themselves, at private schools, and especially, with regard to learning the language and customs of the public sphere, through conversations with their families and friends. This was true also of most of the leading women novelists of the late nineteenth century, such as Sarah Orne Jewett.[32] Jewett's father was a doctor in a small town in Maine, but was well connected by family and education with important people in Boston and Cambridge. It was through these connections that Jewett got her first pieces published in the *Atlantic Monthly*. Her stories in the *Atlantic*, mostly fiction set in Maine, helped to initiate the "local color" movement and brought her encouragement to write novels. Her fiction often dealt (although usually obliquely) with public issues of particular significance for women.[33] If a young woman with literary talent and aspirations were wealthy enough, she could follow a clear trail to the public sphere: finishing school, European travel, fashionable marriage. This was Edith Wharton's path to a literary career; she does not seem to have been impeded in any way by the lack of a college degree.[34]

By the end of the nineteenth century, however, it was becoming customary for upper-class women to attend college and for higher education to be the means of introducing them to new ideas and the possibility of a career in public discourse. Jane Addams, for instance, attended the institution that became Rockford College in Illinois, where she developed many of her writing and organizing skills and a strong desire to do something significant with her life.[35] Although she is best known as an organizer and one of the initiators of the settlement house movement in the United States, Addams was primarily a public intellectual—a writer whose material appeared in most of the quality magazines. A woman did not absolutely have to be well-to-do or even white to find her way into a literary, or at least a journalistic, career through college. The radical journalist Ida B. Wells, the daughter of freed slaves, acquired the skills that made her so effective as an anti-lynching crusader and proponent of racial equality at a college established in her hometown of Holly Springs, Mississippi, by Northern missionaries just after the Civil War.[36] Wells, however, was very unusual, and even she was able to find jobs only with African American newspapers.

Really significant change arrived when women began to graduate in substantial numbers from *public* universities—especially Midwestern state-supported ones—around the turn of the century. It was not that the graduates of state universities necessarily expressed a wider range of political or aesthetic views than did people educated at home or in private colleges. Growth in the number of woman graduates probably did give a boost to movements to increase access to political participation, especially the woman suffrage movement, but it is not clear that the effect was decisive in either defining or leading to the success of such movements. Rather, what happened was that women from a much broader spectrum of social, cultural, and regional backgrounds found the means and encouragement to enter careers in the core public sphere. Women from an equally broad range of backgrounds—in fact, from practically all the categories into which American society was conventionally divided—found their ways though public universities into the body of readers, discussants, and occasional respondents that comprised the basis of the public sphere. University women were exposed to the public conversation under circumstances that encouraged disciplined criticism and comparison among different position. They also learned the language and the body of shared knowledge required for effective participation.

Willa Cather is a classic example of a woman whose career in the public sphere began with attendance at a state university. Raised in a small Midwestern farming community, Cather went to the University of Nebraska, which was decisive in shaping the course of her life.[37] Her family could not have afforded a private college; they had to borrow the money even for the state university tuition from a friend. Cather went to the university intending to prepare for medical school, but once there she found instant success as a writer—with the strong encouragement of the English faculty. She supported herself in college as a freelance reporter and then as a columnist for a newspaper, while simultaneously editing the student literary magazine and writing short stories for national periodicals. Cather went on to become a very successful magazine writer and editor at the turn of the century. In her thirties, she was managing editor of *McClure's Magazine* before turning full-time to writing novels. (*McClure's*, incidentally, pioneered the effort to break holes in the walls of the core public sphere by publishing high-quality writing on all subjects for a general audience understood to be about equally comprising men and women.)[38] Cather, of course, was a spectacular success, but she represents a substantial number of other women from public universities who made careers in the core public sphere in the first half of the twentieth century.[39]

Equally important, however, were the hundreds of thousands of women, their numbers increasing markedly every decade of the twentieth century, who joined the public sphere from state universities as readers and discussants and perhaps as occasional respondents. Most of the readers of this book could, I suspect, think of many women who fit into this category; I will cite my mother, although I could give many other examples. My mother was born into a working-class immigrant family in a copper-mining village in the Upper Peninsula of Michigan during the First World War. She wanted to attend college, but cost was a problem, so she started in the 1930s at the only affordable public college within commuting distance: the Michigan College of Mining and Technology. Michigan Tech was originally established to produce mining engineers, but like practically all such institutions, it had converted itself into a general university-like college with a specialty. Only a few of its students actually expected to become mining engineers—certainly not my mother. She merely wanted to earn credits so that she could transfer to the University of Michigan, which she did. I do not mention this because there is anything unusual about my mother's story, but because there is not. Hundreds of thousands of people had similar experiences in the first half of the twentieth century, a large proportion of them women, as did millions in the second half. The decision to attend college, the necessity of attending a state-supported institution because of cost and proximity and regardless of its official educational specialty, and above all the transfer from one college to another in the confident expectation that studies at the two would be commensurable—all this reflects a pattern that became so common in twentieth-century America that it is easy to forget how remarkable it was in comparison with the immediate past or with what was possible in other countries. (People today making educational policy and obsessing about university "completion rates" certainly do forget the importance of transferability and miss the point that it is not necessarily a bad thing if a public university gives people the capacity to transfer elsewhere, even if it cuts into the number of graduates of the original institution.)

To finish the example briefly: in her later career as a nurse and nursing administrator and during the years that she spent at home raising her children, my mother was *by choice* a full-fledged member of the core public sphere. (My father, although a professional with roughly the same educational and social background as my mother, was not—again, *by choice*.) My mother was not a regular or professional contributor to the published public conversation, but rather a reader, listener, and discussant and someone who understood and used the language in which it was conducted. She subscribed to one "quality" magazine (the *Atlantic,* usually) and regularly read others at the library or subscribed to them for shorter periods. She read books—almost always recent ones—in a wide range of genres. She watched public affairs and "quality" programs on television. My family bought its first television set so that she could watch the Army-McCarthy hearings in 1954. She watched and supported Public Television enthusiastically from the beginning. She did not do much writing for publication, apart from a few letters to editors and pieces in organization newsletters.[40] But she talked about what she had watched and read and thought about with her family, friends, and associates. Conversations with my mother were, in fact, my principal entryway into the core public sphere; college simply refined and extended what talking with my mother had begun. Significantly, though, this had not been my mother's experience. As immigrants with limited formal education who had taught themselves English, her parents had not been participants in the core public sphere and thus could not introduce her to it—even though they read newspapers and took an active part in local community life. The knowledge, skills, and interests that allowed my mother to function in the core public sphere she acquired mainly as an undergraduate in a state college and a state university (even though one of them was in theory a "vocational" institution).

No doubt many of the people who entered the public sphere as my mother did could have done so without the agency of universities, public or otherwise. But the fact remains that the path they took, the most obvious and convenient one, led through the twentieth-century American public university. It made higher education accessible and led to a very large expansion of the core public sphere. It increased the diversity of participation by recruiting readers and discussants from most segments of the social spectrum and providing them with the knowledge and the analytical and linguistic skills needed for full-fledged participation in public conversation. It worked to democratize the core public sphere not with respect to "women" as a single social category that required representation, but with respect to women from the most varied backgrounds.

Although this process was clearly connected with other major changes in the United States, especially the creation of an immense professional middle class,[41] it must be understood as a phenomenon in its own right. The skills and knowledge that qualify a person to be a professional intersect with those of the core public sphere, and at one time, as we have seen, being a professional implied that a person participated in the core public sphere as a matter of course. But the effects of widespread university education on the core public sphere are quite different from its effects on class formation or economic opportunity. They are essentially effects on the way in which the public as a whole—the national community—*functions as a structure,* one within which important matters are discussed and consensus is formed. Social categories (the professional middle class) and individuals seeking individual advancement or security do not act in this way. Moreover, there is the matter of choice. Just because one has a profession (or a university education), it does not follow that one chooses to participate actively or conscientiously in the core public sphere. As we have seen, what has happened to subscription rates for quality journals and newspapers recently suggests that a large proportion of people with professional qualifications may have chosen not to do so.

Democratic diversity is established not only with regard to gender and class, but also race and ethnicity. The effect of public higher education on democratizing the core public sphere with regard to African Americans was rather late in making itself felt, mainly because public colleges and universities open to African Americans in the South were very few in number until the 1930s. But once substantial numbers of African Americans had gone through state institutions, the results were spectacular. One of the most important (although largely unrecognized) sources of the civil rights movement of the 1950s and 1960s was public higher education—particularly in its capacity of bringing people into the public sphere.

As we have seen, the core public sphere in the late nineteenth century and the first decades of the twentieth was not enormously welcoming to educated African Americans. People such as W. E. B. DuBois were not shut out entirely, but they were not able to obtain secure places in it, which forced them to create their own version built around magazines like DuBois's *Crisis*. In addition, discussions of African American higher education in the early years of the twentieth century featured debates that essentially forced the idea of general public universities off to the side. A powerful school of thought, identified with Booker T. Washington and backed by a large part of the white American establishment interested in the subject, focused on vocational education on the model of Tuskeegee Institute and denigrated standard liberal arts-based university education. The principal alternative view, enunciated by DuBois, was mainly concerned with the education of an elite, mostly at high-quality private colleges such as Morehouse and Howard.[42] In the North and the West, African Americans entered standard public universities in small but increasing numbers in the first quarter of the twentieth century. In the South, however, the states for the most part supported a small number of institutions that originally started out teaching trades or preparing school teachers. Most degree-granting colleges for African Americans in the South were small denominational institutions whose standards were, with only a few exceptions, deprecated by national educational commentators.

The situation began to change substantially in the 1920s and 1930s. Around 1900, there had been only about 3,900 African Americans enrolled in any kind of postsecondary educational program anywhere in the country. In 1917, there were about 2,500 students in black colleges in the South, only 12 of them attending state institutions. By 1935 there were 12,600 African American students in public colleges in the South alone and about 17,000 in private colleges throughout the United States. Both the total number of black students and the proportion of them in state-supported institutions continued to grow thereafter.[43] Public higher education was producing neither Washington's vocationally trained farmers and shopkeepers nor DuBois's elite, but rather a version of the same range of educated people that it was producing among the rest of the population.

Many African American graduates intended to enter professions, most of which were segregated de facto. But segregation did not prevent graduates from joining the public sphere as readers of newspapers, magazines, and quality journals; as educated people acquainted with the language and usages of public discourse; and as discussants in the great public conversations of the twentieth century. The group that did these things comprised a minority of African Americans, but by the 1950s, it was not a tiny minority. The descendants of DuBois's "talented tenth," the graduates of Morehouse, Fisk, and Howard (and Harvard and Northern divinity schools) continued to provide the most prominent leadership for African Americans in general, but they no longer constituted an isolated elite. African Americans educated in public and in supposedly "marginal" private colleges, together with others who entered public discussion through churches and radical political organizations and labor unions, were ready to address the larger public sphere in a way that had not been possible before.

The culmination of these developments was the Civil Rights movement of the 1960s. The movement was preeminently a phenomenon of the public sphere. College students played a central role, pushing a more traditional African American leadership into a series of confrontations with white-dominated institutions in the South calculated (with great accuracy) to attract the attention of the national media and to win the high ground in public debate outside the South. Colleges and universities provided African American students with the tools they needed to accomplish all this: socialization in public discourse, entry into the broad public conversation, and knowledge of the language and issues of the core public sphere. Many of these colleges were private African American institutions, but the majority of the students, whether black or white, engaged in the Civil Rights movement in the early 1960s attended public universities.

The importance of colleges and universities and the particular significance of public ones can be demonstrated quantitatively, thanks to the painstaking work of Raymond Arsenault in assembling biographical information on the freedom riders of 1961—the participants in one of the actions that focused the attention of the national public on civil rights in a favorable way. Of the 230 African American freedom riders Arsenault identified, 119, or 52 percent, were college students in 1961. Of these, 68 (57 percent) attended public institutions—overwhelmingly Southern all-black public colleges and universities—while the rest attended private ones (again, predominantly black Southern colleges, of which at least three no longer exist).[44] In addition, 13 riders not enrolled at public colleges or universities in 1961 attended public institutions (mostly integrated Northern ones) before or after that year. (Of the 204 white riders

Arsenault identified, 46.8 percent were college students in 1961. Sixty percent of those attended public institutions.)

The importance of college students in the early Civil Rights movement is clearly and movingly portrayed by David Halberstam in his book on the Nashville lunch-counter sit-ins of 1960. As one would expect, much of the initial leadership came from students (and faculty) at prestigious colleges such as Fisk, in Nashville, but Halberstam shows how the movement became effective when students at low-prestige denominational colleges and especially at public universities (preeminently Tennessee State, also in Nashville) mobilized themselves and became the core of the movement.[45]

The Student Nonviolent Coordinating Committee (SNCC), at least in its early years, was led by people highly skilled in dealing with the public sphere in all of its versions, from the core to the broad "mass" public.[46] They had acquired these skills for the most part in colleges and universities, most of them state-supported. The fact that organizations such as SNCC were able very quickly to adjust to a multiracial membership in the early 1960s was due largely to the fact that its members—by definition, students—regardless of their family or class or regional backgrounds, shared essentially the same general education and the same knowledge of how to conduct public conversation. By the same token, the later history of SNCC, when it radicalized in 1965 and expelled its white members, resulted to a considerable extent from the perception that there were limits to the extent that influencing the core public sphere could change American politics and American life.[47] Nevertheless, what the Civil Rights movement did achieve owed a great deal to the way in which American public higher education had worked to democratize the public sphere, even if the process of democratization was incomplete.

Ever since the mid-nineteenth century, one of the principal stated tasks of public education in the United States has been the "Americanization" of immigrants: particularly non-English-speakers and people whose religious and cultural practices seemed to be substantially outside the mainstream of the nation.[48] In the nineteenth century, this strategy focused on elementary and secondary schools. Higher education was supposed to be above such processes. University students were assumed to be "Americanized" already. By the twentieth century, however, higher education had become the standard pathway for immigrants, and especially for their children, to obtain full individual standing in the upper part of the American mainstream—in the professions, in the middle- to high-income groups, and in the public conversation. They did not attend college to be American*ized,* but to become *full-fledged* Americans. Practically every substantial immigrant group created its own private universities and colleges, directly or indirectly catering to the desire to achieve the complicated aim of integration into the economic and political life of America while retaining cultural distinctiveness. But the large majority of first-generation children of immigrants attended public colleges and universities—many because of financial reasons and some because they did not want to be exclusively identified with a distinct nationality.

Patterns varied, in part because of differences in state and municipal educational policies. In Massachusetts, for example, with a large immigrant population that was primarily Catholic and an array of strong private universities, the state did very little before the mid-twentieth century to provide higher education to anyone except people intending to be teachers, farmers, or merchant marine officers.[49] Catholic institutions and private universities filled the gap (or attempted to do so).

In New York City, on the other hand, a municipal initiative directly and deliberately addressed itself to providing a high-quality university education to the brightest members of low- and middle-income, immigrant, and minority groups. The city sponsored City College of New York and other urban colleges that eventually constituted the City University of New York system. The initiative was connected with New York City's development of an advanced level of first-class public high schools. Not only were the high schools free of tuition charges, but so, too, was CCNY—the catch being that you had to pass a rigorous screening process to get in. The conception was thus strongly meritocratic, which meant that it worked well for people from backgrounds that accorded professional and academic success great value, and probably less well for others. Both initiatives are famous for the way in which they brought the children of recent immigrants into positions in which they could join the very highest levels of American intellectual, artistic, and public life.[50] A significant part of the "Jewish Renaissance" of the twentieth century was created by the children of Russian or Eastern European immigrants who went through the public schools and universities of New York City. Similar successes were scored by members of other immigrant groups.

The examples that have just been discussed all support a crucial general point: not only has public higher education performed the wide range of functions that we usually ascribe to it—producing most of the American professional middle class, providing the knowledge base of a modern "super-skilled" labor force, and preparing citizens for civic roles—but it has also strongly affected the public sphere as a vital element of American society. Public higher education has worked to democratize the core public sphere—to make access to it available to people in almost the entire range of categories into which society can be divided, to build a broad audience for the media of the core public sphere, and to reduce its inherent elitism. Public universities have not, obviously, achieved complete success in these respects, but what they have accomplished is very impressive. One of the ways in which they should be able to make a case for improved state support would be to show this and to show also that much work remains to be done. That work needs to be done is obvious: the core public sphere is in trouble, and part of the reason is that people with college educations appear to be choosing in large numbers not to participate in it. Public higher education is, as we shall see, in a unique position to help to resolve parts of both the fiscal crisis and the performance crisis of the core public sphere, and at the same time to demonstrate its own indispensability.

Why has it not already done so? One reason is that the importance of state-supported universities to public discourse and the public sphere has not been fully recognized at any time in nearly the past century, in part because that function is not acknowledged by the "academic ideology" described in an earlier chapter. Another reason, at least equally important, is that the nature and importance of the core public sphere itself have been occluded—that is, partly hidden, obscured, although not completely ignored—for at least as long. These are the subjects of the next chapter.

CHAPTER 6

Occlusion and Its Consequences

We begin this chapter with a paradox. In the first half of the twentieth century, the core public sphere functioned as a central, vitally important feature of American society and culture. While its role was not exactly what its creators had hoped it would become in the nineteenth century, it was reasonably close. The core public sphere was one of the most important factors that made the modern United States what it was. But it had no name; even more, its existence was regularly left out of consideration in serious discussions about what "the public" meant—discussions that took place almost entirely within the core public sphere. Some (far from all) of the problems of the contemporary public sphere stem from this circumstance. To understand this—both the nature and importance of the core public sphere and the fact of its conceptual occlusion—we will look at a well-known debate over the existence and nature of the "public" in America that was initiated by two books that Walter Lippmann wrote in the 1920s, to which John Dewey published a response in 1927. Lippmann and Dewey between them represent the pinnacle of the core public sphere in the first half of the twentieth century (in Lippmann's case, the first two-thirds). Their careers show us the core public sphere in operation. Their positions on the "public" show us the extent to which the core public sphere was occluded, even in the eyes of two of its most distinguished members.

Walter Lippmann, John Dewey, and the Core Public Sphere

Walter Lippmann (1889–1974) was for half a century the best-known public intellectual in the United States, and reputedly the most influential.[1] Lippmann came from a wealthy New York family. Hardworking, extremely intelligent, and gifted with an admirable literary style, he moved quickly into journalism after graduating from Harvard in 1910. In 1914, Herbert Croly hired him to be a member of the founding staff of the *New Republic,* the liberal but not-too-radical weekly magazine that almost immediately became, for a while, the central vehicle for public discussion of important issues in the United States and has remained a fixture of the core public sphere ever since.[2] So great and immediate was the *New Republic*'s success that by 1916, the Wilson administration was treating it as a significant medium of access to liberal opinion and gave high priority to communicating directly with its editors and principal writers—especially Lippmann. Lippmann was even granted a rare extensive interview with Wilson himself.[3] This was the public sphere working as people like Godkin and Henry Adams had envisioned it, and the 27-year-old Lippmann was at the center of it.

American entry into the First World War initiated a major split in the public sphere. Writers, editors, magazines, and academics had previously lined up for or against entry, and now they lined up for or against the Wilson administration's increasingly severe crackdown on dissent. Lippmann and the *New Republic* had come around to supporting intervention well before it occurred, and both called for the vigorous prosecution of the war.[4] Lippmann joined the government itself, first as an assistant to the Secretary of War and then as secretary to "The Inquiry"—a (more or less) secret committee of academics, publicists, and smart people in New York charged with developing ideas for policy. In this capacity, Lippmann helped to draft the text of Wilson's Fourteen Points. He subsequently received an Army commission to serve as a propagandist in France, where he assisted Colonel House in negotiating with the British and French over the terms of the armistice.[5]

In 1919, however, Lippmann's time at the center of things came to an abrupt end. He had never exerted much real influence on policy, and in any event neither the remains of the Wilson administration nor its successor needed the kinds of links he had represented. Lippmann himself recognized the limits of his position. He was moreover upset by what seemed to him to be the obvious failings of the Versailles Treaty and by the refusal of the United States to join the League of Nations. Lippmann therefore sought a return to journalism, under circumstances that would allow him to comment on the full spectrum of public affairs without being attached to any particular political faction. He wrote for a wide range of magazines, from the *New Republic* to slicker periodicals with broader circulations such as *Vanity Fair*. This opened up the possibility of working for other outlets with a popular audience. The importance of connecting the world in which he operated with that of the average citizen and voter was reinforced for Lippmann when he covered the election of 1920, which disgusted him with the inane discussions of the "issues" that it elicited and the low quality of the presidential candidates of both parties.[6] The election, and the growth of a fashion among social analysts for examining practical democracy with a critical eye, persuaded Lippmann to think long and deeply about the realities of participatory democracy, about the means by which effective government could be afforded in a modern state, and about the ways in which popular behavior was influenced. The result was his 1922 book, *Public Opinion,* which was well received and immediately recognized as a classic. We will return to it shortly.

Just after he finished *Public Opinion* but before it was published, Lippmann accepted a job as assistant editor of the editorial page for the New York *World,* the flagship of the Pulitzer family's newspaper holdings.[7] He had been recruited by Ralph Pulitzer (Joseph's son), who wanted to improve the tone and increase the influence of the *World* without losing readers. The news section remained sensationalist and crusading, whereas under Lippmann, the editorials became more intellectual, evenhanded, and aloof. Lippmann, eventually named executive editor of the whole newspaper, assembled an extremely able editorial team that included the novelist James M. Cain and the historian Alan Nevins. But the paper took on a mildly schizoid character. Lippmann expanded his already extensive reputation and interacted continuously with people in power, listening, giving advice, trying to influence their thinking and in turn sometimes being used by them. His newspaper, however, ran into financial difficulties. It was bought by the Scripps-Howard chain in 1931 and folded into their *Telegraph.* Lippmann moved on to a highly paid position as a regular columnist for the New York *Herald-Tribune,* which he kept until he switched to the *Washington Post* and *Newsweek* at the end of 1962. He had arrived at, and remained in, a location that he perceived as the juncture between the broad public (the readership of the *World,* the *Herald-Tribune,* and *Newsweek*) and those whom he called the "insiders," the people who held political power and made the real decisions in society and the state.

Lippmann's career can serve as a kind of exemplary tour of a significant part of the core American public sphere of the years between about 1910 and the Second World War, and in some respects into the 1960s. It shows among other things the complexity of the intersection between the intellectual, elitist elements of the public sphere and the institutions of the "mass" public. There certainly was a core public sphere and a world of mass media, and Lippmann clearly entered the latter from the former. But the straightforward functional division that had been envisioned just after the Civil War—with the quality press serving as the vehicle through which an intellectual elite would develop the framework of public opinion and the popular press would disseminate the results—had not manifested itself. Instead, popular organs acquired structural connections to the core of the public sphere through editorial pages and people such as Lippmann. Magazines that were oriented neither toward the elite nor toward the audience of the "sensationalist" press also appeared (e.g., *Newsweek* and *Time*); these supported a complex discussion that went beyond mere "dissemination," even if the more detailed, often (but not always) better-informed parts of the public conversation took place in journals such as the *New Republic,* the *Nation,* and *Harper's.* Readership was also more heterogeneous than earlier views of the public had suggested it would be. There was clearly an audience for a wide range of subjects and approaches; somehow (probably to a large extent through public higher education) a public sufficiently well educated and interested in national affairs to support "good," if not always not "the best," magazines and newspapers had appeared.

When we consider Lippmann's public role and career as the consummate "pundit," a fixture at the center of the core public sphere, it makes it all the more surprising that when he turned in the 1920s to analyzing the "public" itself and the nature of public opinion, very little of what he actually did for a living appears. In fact, the core public sphere, as we have been discussing it, does not show up at all. Rather, Lippmann focuses on the role of the *expert*—first in order to emphasize its importance, and later to express his disillusionment about what experts could actually accomplish. Lippmann articulated his views of the public in two books, first in *Public Opinion* and then in *The Phantom Public,* which was published in 1925 after he had worked for the *World* for three years.[8]

In both of his books, Lippmann insists that a "democratic public" and "democratic public opinion" are impossibilities, at least as those terms are usually understood.[9] He explains that this is mainly because of the emergence of the "Great Society" ("great" as "big," not necessarily "admirable"), itself a result of industrialization and the development of the modern nation-state. In its ability to generate material goods and to organize and direct power, the Great Society has outstripped anything else in human history. But the Great Society has introduced a complexity into human social existence that approaches anarchy. The capacity of the modern state for generating power, when combined with its tendencies toward confusion and anarchy, have at their worst produced world war. Even at their best, modern states and the Great Society display a profound inability to establish and follow successful policies. Why is this?

Lippmann's general answer is that the Great Society is simply too complex to be readily understood. Governments "lag;" they are always in a position of having to catch up conceptually with the rapid changes that characterize the Great Society.[10] Lippmann focuses in particular on *democracy* as the factor particularly responsible for the poor performance of politics in the United States and for the lag in adjusting to change. As we have seen, many of the founders of the core public sphere had been critical of democracy, which was one of the reasons they had given for constructing a space for intelligent and informed discourse. Lippmann, however, directs attention not to particular deficiencies of democratic practice, but rather to what he regarded as the flaws in the fundamental theory of democracy in general. Democratic theory requires people to handle tasks that, simply as conscious humans, they cannot perform.[11] Citizens are required to be "omnicompetent:" to be able to understand and make complicated judgments about all the significant issues facing the polity, and to be capable of manifesting this understanding and these judgments in the votes they cast for public officials or on specific issues. This is impossible, even for the most intelligent and interested.

The "omnicompetent citizen" is almost, although not quite, a straw man. Lippmann gives the names of very few "democratic theorists," and none of the ones he cites actually levies such a requirement on the citizenry.[12] He uses the idea, however, to make two entirely valid points. First, it is absurd to believe that the act of conducting an election and counting the votes permits the manifestation of a comprehensive "public opinion."[13] However one conceives of public opinion, it is far too complex to be equated with the result of an election of candidates. Second, standard conceptions of democracy do assume that the average citizen takes a continuous and lively interest in public affairs and does a fair amount of intellectual work as a result of that interest. Clearly, a high proportion of the citizens of democratic countries are willing to do no such thing.[14] The "omnicompetent citizen" is an exaggeration, but the problems of democracy that it suggests are very real. The core public sphere is, in part, a means of dealing with these problems. But Lippmann does not see it that way. In fact, he does not see the public sphere at all.

In *Public Opinion,* Lippmann denies that the citizens are collectively capable of forming a reasoned "public opinion." Instead, what passes for public opinion is a set of beliefs assembled from prejudices and fears shared by a wide spectrum of the general population and constructed around vivid images by the media and by politicians in order to allow "insiders" to retain control of decision-making authority.[15] (Lippmann adapted the word "stereotype" to its current use in order to designate one category of these prejudices.) The logical relationship between what those who control the media and the political parties actually want to achieve and the "public opinion" that they work to create is tenuous at best. Political decisions reflect the interests of political and economic insiders, who employ symbolic actions framed in terms of the prejudices of the public to create the illusion that public opinion has prevailed. Public opinion is thus incapable of acting as the check on the state that liberals have traditionally represented it as being. At the same time, there is no inherent characteristic of the "insiders" that insures that what they do makes sense, either. Unless some structure exists to provide them with information and rational policy options, they are only marginally more able than the average citizen to acquire and apply the knowledge necessary to deal with the modern world. Moreover, their interests are partial, depending on the business or class to which they belong. They are seldom, by themselves, in a position to see the big picture.[16] For these and other reasons, policies that emanate from a democracy are frequently unsuccessful, or worse.

Can anything be done about the weakness of democratic government (or indeed, any kind of government) in confronting modernity? In *Public Opinion*, Lippmann responds with a qualified "yes." The key to the answer lies with *experts* and with altering conceptions of the democratic process to emphasize expertise.[17] The realities of the Great Society are so complex that only trained experts can understand them. Lippmann does not quite say that the experts should actually govern, although he comes close. He calls for the expansion of the boards of experts that are already proliferating in government—as "democratic practice" moves ahead of its lagging theory. Such boards conduct scientific research in the technical areas in which they are competent, collate the research of others, and make objective recommendations to the government.

Lippmann's focus on expertise was not, as we have seen, original or unique. In the late nineteenth and early twentieth centuries, the idea that universities primarily existed to produce experts became very widespread, not just in the United States but throughout much of the rest of the world as well. During the same period, the notion that many of society's problems could be solved if substantial power were turned over to technocratic experts had many supporters on both sides of the Atlantic.

This view represented a substantial difference in emphasis from that of most of the initial founders of the core public sphere. The latter had no objection to professionally trained experts, but they had placed expertise in a conceptual framework of general public discussion among intelligent, educated people. Experts claiming authority solely on the basis of their expertise, establishing policy without reference to public discussion, were not originally part of the picture.

The problem is that without substantial authority and the backing of interests, experts have a great deal of difficulty being heard—except in the arena of public discourse, which Lippmann does not consider. Lippmann subsequently recognized the problem of authority, and it caused him to change his mind just after the publication of *Public Opinion*. In *The Phantom Public*, the people who matter, the ones who exercise power, are the "insiders," whom the experts advise.[18] Lippmann no longer suggests that the power can be shared with experts; the insiders have no reason to share it. All experts can do is to help the insiders develop and choose options. And whereas Lippmann had earlier suggested that the experts, because they practice "science," would be able to give objective, dispassionate advice, now he admits that this is close to impossible. No one is able to be completely objective about human affairs, because everyone is human and therefore has a particular perspective that cannot, in the end, be avoided. The best that can be done is to narrow the range within which biased perspectives operate (mainly through professional ethics and the behavioral standards of science).

But what about democracy? Lippmann ends *Public Opinion* optimistically, with an argument seemingly inconsistent with what he had said earlier. He states that, regardless of the defects that democracy displays, he retains faith in the possibility that reason will prevail in society.[19] If citizens cannot handle the difficult questions that only experts can really understand, they are at least able to judge whether or not the general tendencies of the state are in the right direction and whether or not those running the government are trustworthy. And possibly over time, wider segments of the population can acquire (perhaps through education?) a greater capacity for comprehending the complexities of modern issues. In *The Phantom Public*, there is no such inconsistency. After a quick survey of civics textbooks, Lippmann rules out the possibility that "civic education" in the schools can even begin to equip ordinary people to deal with the complexities of the modern world.[20] The images that fill the minds of citizens and that constitute the reality of public opinion cannot be eliminated. Even the insiders are often prisoners of mental images as well as their own interests. Experts, themselves not free of apparitions and interests, can bring knowledge and science to bear only occasionally. Democracy? Once again, if things are going reasonably well, citizens will generally have some idea that they are and will vote to keep the present insiders in power.[21] If not, they can (perhaps) toss them out.

The Phantom Public was as widely discussed but not as well received as *Public Opinion,* probably because of its unrelieved pessimism. In both books, Lippmann had presented an argument before a literate public of which, despite the books' titles and subject, he had taken no apparent account in the books themselves.[22] He had stirred up interest and a certain amount of controversy within the core public sphere, which was also conspicuously absent from his analysis. The best-known response came from another prominent public intellectual, John Dewey, in a book entitled *The Public and Its Problems.* Dewey actually refers to Lippmann by name only once (favorably, for having framed the subject he was discussing).[23] Nevertheless, Dewey's book is very clearly a reply to Lippmann.

By the 1920s, Dewey (1859-1952) had long been one of the most renowned thinkers in America. An academic philosopher, he had been an early Johns Hopkins Ph.D. and had taught at Michigan, Chicago, and Columbia. Dewey was well known as the successor to William James as the chief advocate of "pragmatism" in philosophy (a term both disliked) and even more famous as a theorist of democratic education. Most of his early publications had been on philosophy, teaching, and social reform. Although he was a generation older than Lippmann, Dewey had begun to take an active part in the discussion of major issues (of all sorts) in the quality national journals at about the same time that Lippmann had. Dewey represents, in other words, another part of the core public sphere: the academic side that was, as we have seen, becoming increasingly central to its operations. If anyone *should* have been aware of the relationship between universities and the public sphere, it was Dewey.[24]

Much of Dewey's response to Lippmann rests on a complex argument about the meanings of "public" and "private" and the distinction between them.[25] In brief outline, it goes this way: Humans must act in conjunction with one another for their actions to have effect and meaning. Transactions among particular people that have consequences that affect only those people are "private"; those that affect other people, however indirectly, are "public" because the other people have an interest in the consequences of the transaction and possibly a need to influence it. Dewey's illustration of a transaction is a "conversation," an exchange of words. All transactions are inherently social; the abstract, autonomous "individual" of traditional liberal theory has little connection with a reality composed of people who have to interact, who must hold conversations with each other. From this analysis, Dewey derives a definition of the collective "public": "The public consists of all those who are affected by the indirect consequences of transactions to such an extent that it is deemed necessary to have those consequences systematically cared for. Officials are those who look out for and take care of the interests thus affected."[26] The state (the "officials") is the product of this recognition. Dewey refuses to acknowledge a clear boundary between the state and the public. The public produces the state directly; "officials" include those who merely vote or otherwise take part in the processes of decision making. Dewey thus implies that any attempt to construct a categorical distinction between public and private, between state and citizen, or (as Lippmann does) between expert and layperson and between "insider" and "outsider," misconstrues the relationships that lie at the basis of political life. Precisely because democracies insist least on such distinctions, they are the political systems most consistent with the reality of the public.

Or rather "publics." Different publics exist simultaneously, although in order to form an effective state, they must share common features. The modern "public" is a complex network of associations that continually changes as the conversations at its center change.[27] Moreover, Dewey's public does not relate to the state contractually. It *is* the state. Open discussion without restriction is essential, not only so that the public can manifest itself but also so that it can, as a whole, supervise and criticize the performance of its own members in their capacities as officials. Dewey's view in these respects thus differs markedly both from traditional liberalism and from Lippmann's analysis of the public.

Dewey does, however, agree substantially with Lippmann with regard to the *actual performance* of the contemporary public. He says that that American democracy "developed out of genuine community life, that is, association in local and small centers where industry was mainly agricultural and where production was carried on mainly with hand tools."[28] The conversations that constituted the public were local ones, directly linked to a system of local, democratic autonomy. But the massive changes of the nineteenth century (especially industrialization and the emergence of the United States as a centralized nation-state) made that system untenable. The interactions of organized associations—parties, interest groups, and so forth—replaced the face-to-face conversations that once constituted the public. According to Dewey, these developments have produced a state based on individual liberty that is vastly more extensive than anyone in the past could have dreamed, but they have also "eclipsed" the public. The public is not, as it is to Lippmann, inherently incompetent to deal with the modern world; it is rather "lost" and cannot "find itself."[29] Because face-to-face conversation can no longer be the mode in which the public operates and no effective replacement has been devised, the essential public cannot constitute itself. Instead of rational discussion among members of communities, there are images and meaningless, symbolic issues created by parties and interest groups trying to elicit commonalities of emotional reaction. Political actions that follow from the issues are irrelevant to the most significant aspects of reality. Citizens are left confused and inarticulate.

The significant point for our purposes is that Dewey does not mention two of the structures of the core public sphere within which he himself operated: the network of quality periodicals and the modern American university.[30] (He does refer to special-interest associations, which are essential to his view of how plural publics are constituted, but he does so only in very general terms.) Although he is certainly in a position to articulate a notion of public discourse in which the core public sphere plays a central role, Dewey does not do so. He presents instead a romantic vision of premodern, preurban localized democracy and moves from that to the deficiencies of a world in which the vision no longer matches reality (assuming that it ever did). Like Lippmann, Dewey ignores the actual framework of institutional public conversation within which he is making his argument.

Dewey also discusses Lippmann's "experts."[31] He accepts them as a necessary result of modernity. He does not, however, think that technocratic elites can govern in any real sense of the term, nor can they, by themselves, form public opinion. The most important aspects of politics concern the shaping of the goals of policy, the formulation of issues, and the posing of questions of value that suffuse the fundamental actions of government. These aspects must necessarily reflect the interests of the plural publics that compose the public at large. Technical experts can carry no more weight in this regard than any other set of informed citizens. They cannot *form* public opinion, because only participation in the transactions that create an interest in "caring for" indirect outcomes can do that. For the same reason, they cannot govern. There is no substitute for a real public able, through extended conversation about the nature of the transactions in which they are engaged, to establish public opinion. It is through the formation of public opinion that the real meanings of actions that inform politics are established and that adjustments to social change are registered. Dewey also implicitly discounts the ability of Lippmann's "insiders" to govern a modern society. Because their interests are narrow and often just as confused as those of anyone else, they, too, must fail if they attempt to rule without reference to a real public opinion—a public opinion that they can try to shape and can sometimes confuse, but which they ultimately cannot do without.

The final part of Dewey's analysis deals with Lippmann's "Great Society," the product of what we would call "modernization."[32] The Great Society deploys enormously efficient productive and communicative technologies. It has dissolved the local frameworks within which most people in the United States and Europe lived before the mid-nineteenth century and replaced them with national ones. It holds out the possibility of substantial improvement in the human condition. The problem is that there is as yet no Great Community to provide a public for the Great Society. The task, therefore, is to establish such a Great Community, and the way to do that is to create the means for having conversations on the appropriate scale.

Dewey's presentation of the general idea is very attractive, but he is disappointingly vague about the specifics of what the Great Community could be and what could be done to produce it. He discusses instead the subject of experts and the relationship between science and the public, which does little to supply the needed description. In the end, Dewey returns to his general point: community must be based on face-to-face interactions or some effective substitute for them, and these do not take place in the Great Society. He suggests that perhaps technology and a will to accomplish change are part of the answer, but he is no more specific than that.

Dewey would probably have been pleased with the internet revolution—in theory, at any rate. Email, blogging, and web publication could all be represented as means of realizing Dewey's intentions. But he provides no effective responses to the predictable problems of a technological solution to the absence of the Great Community, problems that have arisen in recent years with regard to internet discourse. How, for example, do you identify useful and well-informed contributions among the large volume of others that are neither? This, Dewey says, is a job for the "experts," who are supposed to clarify the difference between "fact" and "opinion." But apart from ignoring the question of what happens when the experts disagree or when large numbers of people choose to disregard them, he does not discuss how the experts are to be produced and identified in the first place. He probably thinks that the answer is obvious: by universities. But he does not explore the role of higher education in the public conversation beyond that (unstated) assumption. In fact, he makes very little effort to discuss the function that education in general is supposed to perform in constructing the Great Community, perhaps because he thought he had already written enough about it in his earlier classic, *Democracy and Education*.[33] In other words, not only does a leading participant in the core public sphere writing in the core public sphere ignore its existence while addressing the subject of "the public," but the twentieth century's greatest philosopher of education pays only passing attention to the role of education in constituting the conversation and the discursive community that he sees as the essence of the public.

Had Dewey deployed a conscious conception of the public sphere, he would have had at least a partial response to some of the problems he identified in "the public." One of the functions of the core public sphere is to formulate issues and to evaluate and discuss expert contributions to their solution. People who thoughtfully read the quality media are not left to assess conflicting or difficult positions by themselves; they do these things mainly by following, and sometimes repeating or taking part in, the discussions and debates that fill the pages of most serious media. The contributions of experts are not disseminated directly to an undifferentiated public but for the most part proceed through the organs of the core public sphere, where they are considered and discussed and criticized. Interested people take college courses or participate in public discussions or (nowadays) watch public television or visit web sites where expertise is queried and contextualized. In *The Public and Its Problems,* Dewey had, by elucidating the central function of conversation in constituting both community and public, established a framework within which the public sphere as we have been discussing it could be readily understood. But he did not make use of what he had created. Why not?

Perhaps Dewey was so concerned with Lippmann's critique of democracy that he focused entirely on the main points of Lippmann's argument, which led him to adopt its general framework and limitations while questioning its inferences. Possibly Dewey actually recognized what we have been calling the core public sphere but did not see it as something entirely consistent with democracy. He had become a contributor to the *New Republic* and other "opinion" journals rather late in his career, and there is evidence that, as a radical democrat, he was uncomfortable with what he saw as their antidemocratic tendencies.[34] He may have believed that his *New Republic* colleagues were speaking mainly to each other and to a limited circle of people who thought as they did—that they were offering themselves (unsuccessfully) as a substitute for a real public. He does not, however, say this in *The Public and Its Problems,* any more than Lippmann makes the offer in his books. The core public sphere is simply not there.

I think that the principal reason in Dewey's case—as in Lippmann's—is that the realities of a functioning structure of informed public discourse had never been reflected in a fully articulated concept in the United States, even when the founders of the core public sphere were deliberately trying to build such a structure in the latter part of the nineteenth century. Thereafter, consciousness of the structure's existence tended to be fragmented into more limited images addressed in disparate discourses: journalism, in a task-specific professionalism; social science, in an idiom of expertise; the image of an intelligentsia centered in New York, aggressively oriented toward culture, aesthetics, and Europe; and higher education. With regard to the last, universities and academia had also developed images, discourses, and forms of structured self-consciousness that occluded their roles in the public sphere. They were doing this, and continued to do it, even as the public universities were emerging as the structures in American society most responsible for democratizing the public sphere. We cannot pursue this particular form of occlusion comprehensively, but we can see how it presented itself in the pronouncements of three of the most prominent figures in American higher education between the 1920s and the 1960s.

Academic Ideology and the Public Sphere

In 1930, Abraham Flexner, the author of the 1910 "Flexner Report" on medical schools and for many years the principal education expert of the Rockefeller Foundation, published a book based on lectures he had given at Oxford that he entitled *Universities. American English German.* When the book reappeared in 1968, it included an introduction by Clark Kerr, president of the University of California and a major figure in shaping the contemporary idea of the "research university." Kerr argued that *Universities* was a classic, but not for the reason that Flexner might have expected. Flexner emphasized what he saw as the decline of higher education since the exciting days of the late nineteenth century when the new style of American university was being created. Flexner wanted to return to the original model, which he called a university suitable for the modern world. He apparently thought that his book would have an effect equivalent to that of his 1910 report. It did not. According to Kerr, Flexner misunderstood the direction in which universities were going. Instead of a program for change, Flexner produced a "valedictory." Nevertheless, Kerr says, what he did in the book had great value. Flexner "preserved for us, in perhaps its purest and most completely reasoned form, the 'idea of a modern university' at a crucial stage in its development."[35] By insisting on the "earlier university's... finest elements"—regard for excellence, the alliance of teaching with research, commitment to intellectual discovery—and by attacking such things as an overemphasis on sports and "absurd topics for the Ph.D.," Flexner issued strictures that "echo down the decades in the hallways and faculty clubs and committee rooms of academe."[36]

Exactly. What is significant about *Universities* is the way in which Flexner articulated an ideology of higher education that had become by 1930 standard for a large part of American academia. Not everyone subscribed to the ideology, which in any event admitted of substantial variation. But in its basic outlines, the pattern of assertions he employed in *Universities* was quite widely accepted by academics in the arts and sciences fields, and so it has remained, regardless of glaring inconsistencies with other aspects of the reality of higher education. There is much that could be said about its virtues and its failings. For our purposes, though, the important point is that the ideology did not make a place for the public sphere as we have been discussing it. It disguised a large part of the relationship between higher education and the core public sphere and denigrated much of the rest.

This can be seen in Flexner's treatment of the relationship between the university and democracy. In the course of explaining that universities must change with social circumstances (although not in the way they actually *have* changed) and must address contemporary problems, Flexner ascribes these problems principally to democracy: "Democracy has dragged in its wake social, economic, educational, and political problems infinitely more perplexing than the relatively simple problem which its credulous crusaders undertook to solve." Nevertheless, democracy is what we have, the product of apparently inevitable social changes to which we must respond. It is better to accept it than to think about replacing it. We must make "adaptations" to democracy. "But adaptations in what ways? Statesmen must invent—not statesmen, fumbling in the dark or living on phrases, but statesmen equipped by disinterested students of society with the knowledge needed for courageous and intelligent action."[37] Where are these students of society to be found who can do democracy's thinking for it? "Men of action" and journalists are not sufficiently disinterested. "About the only available agency is the university. The university must shelter and develop thinkers, experimenters, teachers, and students, who, without responsibility for action, will explore the phenomena of social life and endeavour to understand them."[38]

In one way, Flexner's view reproduces an aspect of the thinking that motivated the founders of the core public sphere and that underlay a great deal of "Progressive" reform: democracy needs help and correction. But there is a major difference. According to most of the founding generation, participants in intelligent public discourse were supposed to be active in the political life of society, or at least they were not supposed to separate themselves from the public competition of ideas. The distance and objectivity that were expected to make for useful contributions to the public conversation—including those of academics—were to be products of the culture of the public sphere and of the convention of "bracketing" the space of public discourse for purposes of argument, not of removal altogether from the space of competition. To someone like E. L. Godkin, setting apart a class of highly educated people to think about things—even socially relevant things—and then to present their conclusions "without responsibility for action" was absurd and dangerous.[39] Godkin described it as creating a priesthood, one singularly devoid of efficacy in the modern world. But that is exactly what Flexner advocates, although he does not use the term "priesthood."

Flexner views universities as places where the highest intellects are assembled and where they engage in thinking about knowledge conserved from the past and in producing new knowledge through research. The research may be, and some of it must be, directed toward the solution of contemporary problems, but the motives for doing the research must be "pure"—that is, researchers should do what they do entirely for its own sake. They should have no other involvement in what they are studying and no personal interest in the outcome of any issues that are to be decided. There must be no constraint, fiscal or otherwise, on their choice of topics to profess or investigate (as long as the topics are not "vocational"). To Flexner, the ideal "university" is a graduate school, consisting of scholars and scientists and their acolytes, the graduate students who will ultimately take their places. There may be some professional schools as well, but the only professional fields that Flexner is willing to accept as appropriate for a modern university are the traditional ones of medicine and law. These are permitted because Flexner claims that they involve the "application of free, resourceful, unhampered intelligence to the solution of problems." Both possess "primarily objective, intellectual, and altruistic purposes" and codes of honor.[40] He rules out all other professions—management, accounting, engineering, even journalism—because they do not have these characteristics (and therefore are not "true" professions.)

These attitudes—the notion of "ivory tower" engagement/disengagement with the world of action, the intellectual elitism of the university, the denigration of most forms of professional education—are still shared to one degree or another by a large proportion of academics, especially in the arts and science fields, although they are seldom articulated quite as openly (or brutally) as Flexner did in *Universities*. In another place, I might take issue with many of them in their own terms. As ideals, they have substantial attractions, especially for faculty with a research orientation. As elements of the reality of higher education (for their very presence in Kerr's "hallways and faculty clubs and committee rooms of academe" makes them a reality), their correspondence to the range of functions that contemporary universities perform is tenuous at best. At one time—between the 1930s and the 1970s—they were widely accepted, at least in part, by groups outside of universities that made educational policy and supplied funding for higher education. Since the beginning of the attacks on higher education that commenced in the 1970s, however, they have more frequently come to be viewed as means by which academics defend their own interests (which is, as we have seen, one of the reasons the academic ideology was articulated in the first place). A significant factor in the crisis of public higher education is the continuing tendency of university faculty to explain their functions in terms of these aspects of the academic ideology, which nonacademics simply do not find as convincing as they did in the years immediately after Flexner published *Universities*. But for present purposes, the important point about the views expressed by Flexner is the way in which they obscure the relationship between the university and the core public sphere. They leave no place for multidimensional interchange between intelligent people inside and outside academia, and they downplay and often disparage the extensive interchange that actually takes place and that is so vital to public discourse.

Like the "experts" in Lippmann's theoretical world, Flexner's ideal academics do not engage in discussion outside their walls, but rather present the results of their objective research to be accepted and implemented—or not. Unlike Lippmann, however, Flexner insists that only research that is motivated by the interest of the researcher in the intellectual problems involved can qualify for attention. How motivation of this sort is to be detected or measured is not clear. This image was, as we have seen, derived from the developing ideology of "objectivity" in American academia in the late nineteenth century (and ultimately from a particular strain of German academic ideology). Flexner, however, carries it a great deal farther than the predecessors he admires so much. Flexner, for instance, has nothing but scorn for journalists and for journalism as a profession. Journalists cannot think or do research of the kind needed for dealing with social problems. Like "men of action," they must have their thinking done for them by objective and disengaged academics. Flexner's particular hero among the university reformers, Daniel Coit Gilman, made claims for university research as a source of objective knowledge to be employed by the state, but he also strongly affirmed the importance of public discourse and listed journalists among the most important of the professionals whom universities were supposed to educate in the interests of society as a whole.[41] This is entirely missing from Flexner's presentation.

Another, more basic and pervasive aspect of Flexner's educational outlook was his belief in rigidly defined categories of people and the necessity of separating as much as possible the kinds of education they receive. Again, Flexner was more extreme in this regard (at least in theory) than many others who thought (and think) along similar lines. There is evidence, for example, that he believed that women and African Americans were not suited to the intellectual rigors of real universities and that it was a mistake to provide them with access to such institutions.[42] But apart from that, Flexner's notion of categories of fitness reflects views still held by many academics and others who make or discuss educational policy, even when they acknowledge (as Flexner did) that deviations from the ideal are required in practice. This aspect of educational ideology impinges on the public sphere in several ways.

To begin at the top, Flexner says that it is not the job of the university to train "practical men, who, faced with the responsibility for action, will do the best they can. Between the student of political and social problems and the journalist, industrialist, merchant, viceroy, member of Parliament or Congress, there is a gap which the university cannot fill."[43] "Practical men" may train themselves or perhaps be prepared by the equivalent of the first two years of an American college, but they do not belong in universities. Flexner probably thinks he is restating Humboldt's idea that universities should not teach "applied" subjects, but he misses a crucial part of Humboldt's conception: that humanistic university education *is* intended, among other things, to prepare people to act and also to fill the "gap" that Flexner identifies. That "gap" is not a gap at all, but the public sphere. In Flexner's view, academic intellectuals and "practical men" simply have different kinds of mind, perhaps different characters, and that is all there is to it. Scholars and men of action neither mix nor, in any meaningful way, even converse. The latter provide nothing of an intellectual nature to the former—just material support. The scholars, as a by-product of research that that they do entirely "for its own sake," provide complete packages of advice to the men of action.

Men of action are not alone in being excluded from universities. Flexner wants generally to leave out the people who are not intelligent enough, not sufficiently interested in the life of the mind, not willing enough to dispense with childish sport or concern with material goods. In brief, he wants to debar the average American college student. Some compromise is necessary on financial grounds, but only the intellectual elite should be allowed into real university education.[44] With respect to the masses of people who would not normally attend college at all, most of them should not even clutter the high schools, which should be devoted to preparing students for university and selecting the ones who actually should attend. It is necessary, however, to identify among the masses the brightest prospects and bring *them* into the high schools. This openness to individual talent regardless of social origins, Flexner claims, makes his views entirely democratic in the proper sense of the word. Organizing education so that anyone, regardless of intelligence and aptitude, can participate in any form of education is democracy in its worst sense, and it is what Flexner believes has become the dominant concept in the United States in his time.[45]

Again, Flexner's ideas could be addressed in general terms. He helped to reinforce, although he by no means originated, the idea that there is a single standard of intellectual aptitude—possibly with more than one dimension, but still limited in range—that can be used to separate those who should receive higher education from those who should not. This idea has permeated educational thinking in the United States and elsewhere for many years, even though it does not correspond to either the wide range of types of ability found in any population or the multiple functions that education actually performs, and must perform, in a modern society.[46] But what is important here is the implication of Flexner's thinking for the relationship between education and the public sphere. Not only does Flexner, in discussing the difference between academics and "practical men," provide no space for the core public sphere, but in his view of the appropriate education for the majority of people, he also shows no recognition of any need for them to be able to participate in, or even follow, considerations of policy or any other important public matters. Apparently, they are to be trained to perform their jobs, to be law-abiding citizens, and perhaps (although Flexner does not say so), to be, as Lippmann puts it, conscious of whether national affairs are going well or badly, but not to engage in public discourse. Since public discourse itself appears to consist of university people giving objective advice to "men of action" and the latter (another elite group) deciding whether or not to follow it, there does not seem to be much reason that they should be. The public sphere—whether the core public sphere or the broader field of general public discussion—has again been occluded, at the very least.

A leading figure in higher education who sounded variations on some of Flexner's themes but developed them in different directions between the 1930s and the 1950s was Robert M. Hutchins, president of the University of Chicago from 1929 until 1945 and chancellor of the university until 1951. Hutchins, a combination of *Wunderkind* and *enfant terrible*, was appointed president at Chicago at the age of 30.[47] The timing of his appointment is significant in terms of what was just said about Flexner. One of Flexner's prime examples of the deficiencies of American higher education—its lack of standards, its pandering to commerce, vocational concerns, trivial interests, and football, its lack of a coherent curriculum— was the University of Chicago of the 1920s. Hutchins undertook the task of remaking Chicago at least in part in accordance with Flexner's kind of thinking. In a small book published in 1936, he outlined the changes he wanted to make, not only at Chicago but throughout the United States.[48] The book echoes many of Flexner's ideas and adds several more—the ones for which Hutchins was already becoming famous and which continue to constitute a significant subset of American academic ideology.

The point with which Hutchins begins is similar to one Flexner emphasizes: the curricula of American schools at all levels (but especially in college) are incoherent, mainly because no single, dominant function has been identified for any institutional type. The American college "is partly high school, partly university, partly general, partly special. Frequently it looks like a teacher-training institution. Frequently it looks like nothing at all."[49] Universities contain professional schools and Ph.D. programs, but nobody knows precisely what the purpose of either is. There are redeeming features, but the picture is overall a negative one, based on the criteria of coherence and unitary mission.

This emphasis on unity of purpose is, when one thinks about it, rather peculiar, although it has been repeated by American commentators in higher education down to the present. How many institutions, no matter how successful, have single purposes, and why should it be desirable or necessary that they have them? Yet Flexner and Hutchins both insist on it and make it a major basis for judgment about what should and should not be accepted as part of a university. Why is this? Hutchins does not explain directly, but it is possible to infer his reasons from other parts of his argument. Higher education must be unitary because its main function is to preserve and propagate the truth, and the truth is unitary. Engagement with motives such as "love of money" and with the imperfections and compromises of practical life obscures "the pursuit of truth for its own sake," which is the particular responsibility of universities.[50]

Hutchins was famously an advocate of what he and his associate Mortimer Adler called the "Great Books" approach to higher education. The approach assumes that certain books contain the manifest wisdom of the "Western" world, in the sense that their authors identify the fundamental questions about human life; about existence and the universe; and about the relationships among reason, belief, and morality that must be answered in order to understand the meanings of things. The Great Books also contain the entire range of possible answers. Truth lies not in a single answer, but in examining, discussing and building on all of them, within the structure of the "Western" tradition. Not all ideas are equal. There is a "hierarchy of truths which shows us which are fundamental and which subsidiary, which significant and which not."[51] According to Hutchins, the object of establishing the "hierarchy of truths" is to achieve "real unity" (in ideas, but perhaps also in the community of people who discuss them). The aim of creating unity is achieved by establishing the relative "trueness" of propositions—that is, their positions in the hierarchy. This suggests that it may not really be "truth" (in any common understanding of the word) that is arrived at by discussions within the range of admissible ideas, but rather order.

Although Flexner also wrote about the pursuit of truth, he presented it primarily as open-ended empirical research, with the natural sciences as the model. Hutchins, on the other hand, argues that the only research that should be permitted as part of the "higher learning" at universities is that which deals with "fundamental principles" in metaphysics, social science, and natural science—categories he describes as "exhaustive."[52] In each case, the central activity of "research" is to be reasoning from first principles. Hutchins dismisses most forms of empirical research as mere "data-collecting." He wants to ban them from the university curriculum;

they can be undertaken by research institutes when they can be shown to be socially useful.[53]

Hutchins's greatest influence on contemporary American higher education lay in shaping a substantial part of the discussion of general education. In his book, he presents a model for general education that he suggests should be adopted nationally.[54] The program should focus on providing a framework for understanding and discussing serious issues; it should not include vocational training, nor should it emphasize what Hutchins considered to be nonintellectual exercises such as the learning of languages. (He later changed his mind about the last point.)[55] The last two years of high school and the first two years of college are to be devoted entirely to Hutchins's sort of general education, which will be structured so as to encourage students to become critical thinkers (and thus good citizens) by getting them to read and discuss the great books. They can then go on either to higher education at universities or into jobs. According to Hutchins, this will work for about two-thirds of the people of the right age, the ones who are "literate;" there needs to be no substantial difference between the curricula for the university bound and for the vocationally intended. For the one-third who are not literate, it is still necessary that ways be found to get across the principal elements of general education by means other than reading. Hutchins does not say what these means might be.

Hutchins describes his program as "democratic." Like Flexner, Hutchins criticizes many current attitudes toward American education that he calls "democratic," but he speaks positively about democracy in other respects—as long as it corresponds to his own view of what democracy is.[56] It is not merely the best of a bad lot of competing ideologies, the undesirable aspects of which are to be counterbalanced by the expertise of university researchers. He also has a much less restrictive, and thus in some sense more democratic, idea of the qualifications for university admittance than Flexner does. At its core, Hutchins's "democracy" appears to be a discussion, one that is open to anybody with the appropriate education (but closed to others) and that focuses on the eternal verities: an immense Great Books seminar that debates what the "good life" is, how collective decisions can be made in accordance with rational morality, and why it is important to fight tendencies toward materialism and selfishness in ourselves.[57] The connection between Hutchins's ideas about general education and his notions of political democracy are so strong as to make them essentially extensions of each other. General education, and to some degree higher education, do not so much support democracy as they constitute and model it.

At first glance, it might seem that by describing intelligent conversation as the central feature of democratic political life, Hutchins has fully incorporated the public sphere. This is an illusion. Hutchins's notion of discourse is so narrow and so obsessed with order and intellectual unity that it effectively eliminates the public sphere. His model is an imaginary classical Greek *polis,* where public discussion, framed consciously in terms of a set of shared notions of truth, focuses directly on proposals for state action which are then immediately implemented. There is none of the separation from political action that Habermas emphasizes in the public sphere, nor is there "bracketing" of the conversation in terms of the interests, social standing, and political involvements of the participants. To Hutchins, there is no need for these things. If potential participants in the public conversation can be taught (through general education) to focus on the essentials of what they are talking about and to employ classical models of intelligent discourse, bracketing and separation of spheres will be unnecessary. The participants will have internalized the logical and moral frameworks that will permit them to consider issues objectively and appropriately. In place of the disorder of open dispute bounded only by presumptions of logic and civility, there will be the order imposed by a shared framework of thought and the hierarchy of ideas that education is supposed to produce.

Hutchins's vision has appealed to a not-insignificant number of people who have thought about the relationship between education and democracy. It still exerts considerable influence in discussions of general education in the United States, much of it (I would say) beneficial. It seems to me, however, that the vision is vastly too optimistic about the extent to which a particular form of education can produce voluntary effective consensus among large numbers of people (coerced consensus is another matter), is much too broad in its claims that reasoning in a particular way from the hierarchy of truths revealed by the Great Books can offer resolutions to issues that appear before the public in a complex society (or indeed, even in a simple one), and, perhaps most important, contains no satisfactory answer to the question of who decides which books are "great" and what the hierarchy of truth actually is. These matters could be discussed at considerable length. But again, for present purposes the important point is that Hutchins's contribution to the academic ideology did nothing to reveal or develop the place of higher education in the American public sphere and in some measure helped to obscure it.

We are not quite finished with Hutchins. In the 1940s, he chaired a highly publicized commission of prominent intellectuals formed to investigate the state of the public media.[58] The commission considered a wide range of what its members perceived as deficiencies in the press: tendencies toward control by monopolists who used the right of ownership to shape news and editorial opinions in their own interests; shoddy reporting, at variance with professional codes of journalistic ethics; deliberate withholding of the news; sensationalism. The report of the commission, issued in 1947, has generally been called the "Hutchins Report." It is still frequently cited and has had a substantial effect on academic theorists of journalism who have shaped concepts of "public journalism" and the "social responsibility" of the media.[59] It codified a number of propositions about the role of newspapers in a liberal democratic society that had been articulated for some time, placing them in a framework in which, at least in theory, the public obligations of the press were held to trump the interests of owners.

The Hutchins Report elicited a great deal of criticism in the 1940s from journalists and newspaper owners. The issues it raised have been discussed recently in the context of the crisis of public discourse. One of the most interesting elements of the report is its statement about what the press is obliged to provide to the public in fulfillment of its social responsibilities: "a truthful, comprehensive, and intelligent account of the day's events in a context which gives them meaning;... a forum for the exchange of comment and criticism;... the projection of a representative picture of the constituent groups in the society;... the presentation and clarification of the goals and values of the society;... full access to the day's intelligence."[60] The second-to-last item appears to be pure Hutchins. The rest are essentially what we have come to expect from newspapers and magazines, especially those that advertise themselves as being of high quality. When they do not deliver these things, we criticize them.

From our perspective, the important item in the list is the second: news media are supposed to provide a forum for comment and criticism. They are, in other words, required to play their part in the public sphere. The report was referring mainly to such things as letters to the editor and regular columns by professional commentators, as well as criticism of films, art, and so forth. All had been standard in leading newspapers for some time by the 1940s. Lippmann's columns had become the models for professional commentary; their tone of judicious nonpartisanship paralleled the emphasis on objectivity that had also become standard in news journalism. The Hutchins Report was, in other words, registering a reality by implying that the functions of finding, reporting and interpreting the news was intimately linked to public discourse. The report devoted only a small amount of space to discussing this item, but at least it was formally noted, which was better than Lippmann—a professional journalist—had managed when discussing the "public" in the 1920s or than Hutchins had done by himself when considering the relationship of higher education to society.

To round out the discussion, we can look briefly at James Bryant Conant. Conant is probably best known for having been president of Harvard from 1933 to 1953, during the period in which Harvard converted itself into the model of the modern research university while at the same time preserving its reputation for leadership in many other aspects of higher education. He was also a very successful research chemist and historian of science, an organizer of the Manhattan Project and the postwar federal nuclear energy program, and United States High Commissioner in West Germany.[61] All of these aspects of Conant's active life were relevant to the one that concerns us here: Conant as a leading postwar contributor to the ideology of the modern American university. He published his broadest treatment of his view of American universities and colleges in 1956 in a short book called *The Citadel of Learning*.[62]

The title of the book refers to the context in which Conant places it: the Cold War. Contrasting ideas of higher education in the West with those imposed in the Soviet bloc, Conant focuses on freedom of speech, encouragement of discussion, and tolerance of dissent as the key features of the former that distinguish them from the latter. The university in the West, he wrote, is a citadel defending these things against attempts to impose uniformity. Because what the universities do is central to the meaning of Western civilization, it is vitally important that they be supported.[63] This is clearly not just an attack on Stalinism, but also by implication a criticism of McCarthyism and its threat to free speech. But, presumably because his topic is universities and not the public in general, Conant moves immediately to the special role of the university in promoting the creation of new knowledge for the benefit of society. His point is that good science (like good scholarship in any field) requires toleration, argument, and dissent. There is obviously nothing objectionable in this, but Conant avoids addressing the role of the university in participating in broad public discussion except as a provider of the results of research. I have no doubt that if he were writing more generally about the public, Conant would agree that dissent and argument must be protected there as well (although probably not for the same reason). But he does not say that the university has a particular role in conducting or supporting public discourse. Once again, the connection between higher education and the public sphere is not so much denied as occluded, passed over, represented only partially.[64]

Conant then turns to a comparison between American and European higher education. Here, Conant appears to be attacking people like Flexner who "fail to recognize the mutation that education has undergone in the last century in America" and whose "criticism of our schools and colleges is often not only harsh but unrealistic, and consequently not constructive."[65] He seems to be delivering a ringing defense of the kind of higher education system made possible by public universities (nonelitist, serving a large number of students with varied backgrounds and aims) as something peculiarly appropriate in America. (This part of the book was adapted from a speech Conant gave at Michigan State's centennial celebration.) But then he asks, "Does the American tradition in education... stand in need of modification to meet the challenge of our new world, the constructed globe of the mid-twentieth century?"[66] It turns out that it does, that what he has just said refers to the United States in the immediate past. Now that America is the leader of the Free World, things are different.

Conant gives a long list of needed changes, most of which quickly became clichés in the discourse of American education. The curricula of high schools and colleges should be structured so as to prepare at least some students for overseas service, among other things by emphasizing language instruction. Students need to understand the historical contexts in which policy has been made in the past. America must produce more scientists and engineers. A system should be constructed for identifying scientific talent early so that enough qualified students will be ready for intensive scientific study of science by the time they get to the university.[67] Conant's focus is on the relatively limited number of students on whom the principal burdens of intelligent national service will fall, although he carefully avoids calling these people an "elite." Some of his recommendations, however, address the broad range of students and deal with preparing them to be members of the public.

Along with teaching students "the nature of the opposition between the doctrines we hold dear and those of the totalitarian Communist regime," Conant wants undergraduate general education programs to teach citizens how to "evaluate" the statements of experts (especially scientists), mainly by providing them with information so that they can, for instance, "without themselves being prepared to appraise a foreign nation, realize some of the complexities and difficulties of the task."[68] We might expect Conant now to discuss the nature of public discourse, its relationship to education, its role in constituting the free society that must be protected against Communism. But he does not. The context in which "the proverbial man in the street," equipped with a general education, is supposed to evaluate and recognize complexity is not described. Conant does not seem to be saying that the general public is involved directly in decisions about policy or that most of its members can contribute to a discussion that will influence such decisions. He gives no attention to the relationship of universities to the discussion that takes place in newspapers and magazines. Conant apparently wants members of the generally educated part of the public to understand enough about, for example, science that they will accept the correctness of what they are told by scientists— not all scientists, but the ones who present the consensus on a particular topic. The public is not being invited to take part in a discussion of the disputes among experts, which Conant has already sited within the walls of academia. The public is to be aware of the disputes and to await the outcome, assenting to the conclusions of the experts in part because it understands how difficult their task it. In other words, Conant is not thinking of the public sphere, whether narrowly or broadly conceived. Again, the core public sphere has been overlooked.

Toward the end of *The Citadel of Learning*, Conant gets to the two topics with regard to which he exercised his greatest influence: meritocratic education and the research university.[69] In my opinion, the aspects of contemporary academic ideology and of public expectations about universities that have developed from the views Conant (and others) expressed on these topics at midcentury constitute major factors in the present crisis of American higher education. To discuss them, however, would take us far beyond the scope of this book. It must be sufficient here to say that in his discussion of neither of these topics does the core public sphere appear in any significant way.

Conant, Hutchins and Flexner present views of what American higher education is supposed to be about that were current in the middle years of the twentieth century and which have continued to inform a large part of the ideology of American academics ever since—and a good deal of the framework within which universities were understood for several decades by the part of the public that most influenced the assignment of resources to support of higher education. In the former instance—the academic ideology—these views were attached to the set of concepts used by tenured university faculty to justify the interesting, respected, reasonably well-paid, secure, and essentially autonomous career pattern they had created for themselves. It is not surprising that they should be less than eager to subject the academic ideology as a whole to criticism or to entertain the idea that substantial changes might be called for. Not everyone in higher education has accepted the basic tenets of the ideology. Alternatives based on radically different assumptions have been propounded and, occasionally and very temporarily, realized. It is possible that the current flood of "online universities" will effect a permanent transformation, although considering their lack of prestige and their tendency (although not in every case) to replicate the most unimaginative parts of the curricula of "standard" universities, I doubt it.

Many of the values that the academic ideology expresses are unobjectionable and some are admirable, although as I have hinted, some are in my opinion considerably less admirable than others. Most of my criticisms are not directly relevant to this book. But from the perspective of the contemporary crises of the core public sphere and public university, the continued prevalence of the academic ideology poses two problems. One is circumstantial: the part of the public that influences decisions about education is no longer as easily convinced as it was fifty, even thirty, years ago that the main tenets of the ideology are valid, which has caused the defenders of universities (especially the public ones) to emphasize instead the economic functions of higher education and to downplay its other contributions. As we have seen, not only has this created serious conceptual dislocation, but it is ultimately a self-defeating strategy both for the universities and for the nation. The second problem is internal to the academic ideology itself. As we have seen, the ideology does not accurately represent some of the principal functions and achievements of American higher education, especially its role as a vital part of the core public sphere. This has prevented universities from fulfilling the obligations that follow from this role as well as they might have done, thus contributing to the crisis of performance of the core public sphere. It has also prevented the state universities from making as strong a case as they could for maintaining and increasing support from public sources. There is a very good reason for public higher education that is not articulated nearly as often and as forcefully as it should be, one that lies in the need for an effective, democratic public sphere and not solely in the imperatives of a "knowledge-based economy."

CHAPTER 7

What Should Be Done?

In this chapter, we arrive at the crucial questions to which the other chapters have been leading. If, as has been argued, the American core public sphere plays a necessary role in the United States and the world, and if, as has been shown, the core public sphere is in crisis—in part because of inherent weaknesses and because of the obscurity of general comprehension of its nature and functions—what can we do to resolve the crisis in a desirable way? If, as has been demonstrated, public higher education plays a definitive role in American society and has performed an essential one with regard to the core public sphere, what can we do to resolve *its* crisis satisfactorily? And if the core public sphere and public higher education (indeed, higher education in general) are in fact so closely connected, how can we shape responses to the problems of colleges and universities in such a way as to support desirable changes in the core public sphere as well?

It has been suggested that many of the problems of public discourse in the United States might be solved by the proper use of modern information technology. Recently, Al Gore has proposed that the internet and the World Wide Web have made it possible to create a democratic society of well-educated citizens conversing with each other on a regular basis, sorting out the vast variety of information needed to create an informed opinion and seeking expert opinion when necessary.[1] Perhaps such a society would make it possible to do without the core public sphere altogether.

It seems to me that the internet may help to change the structure and practices of the discursive public (in fact, it already has), but it cannot do so in the absence of something like the core public sphere. For one thing, information technology does not by itself obviate the need for a recognized mechanism for creating some degree of consensus among people who know what they are talking about, for signifying who these people are, for ensuring that the consensus is established openly and transparently and is itself subjected to continuing intelligent examination. This is one of the functions partly performed by universities, which certify the expertise of their graduates (especially those with doctoral degrees) and the members of their faculties. It is also one of the functions of the "quality" newspapers and magazines. We have seen that the function is often not well performed, that expert consensus is sometimes dead wrong and reflective merely of hierarchies of prestige and power, that the model of experts providing the "truth" of a subject to a public that is supposed to accept it often undermines and obscures the working of the discursive public. At the same time, though, it does not appear to be possible to do without the function. It is true that several internet magazines and blogs have attained authoritative status, but that simply means that they have joined the inner circle of the core public sphere, not replaced the thing itself. They differ very little from the quality print magazines except in the mode of their transmission and in the fact that they are not as closely tied to a publication schedule. Most of them share with many of their print counterparts a continuing revenue problem that being on the internet has in no way solved. The problems of trying to create a consensus just on matters of fact and first-level interpretation without expert certification of the credibility of sources can be seen by looking through entries in *Wikipedia*.

Wikipedia is a wonderful innovation, with no real precedent, something impossible to imagine without contemporary information technology. It also demonstrates the extent of informed interest in the widest conceivable array of topics and the willingness (admittedly sometimes self-interested) of individuals to take part in a collective intellectual enterprise. It is also enormously useful. I consult it almost daily. Using *Wikipedia* effectively requires one to read the articles critically, knowing how they are written and understanding that they will often necessarily contain inconsistencies, contradictions, and occasionally self-serving falsifications. One should read *any* source critically (which means, incidentally, that one should be taught how to do so—not just "facts," but how to analyze statements that purport to represent facts). But with *Wikipedia* there is no effective way to tell, without considerable research, whether any particular line in an entry was put there by someone who knew what he or she was talking about. Sources are cited, but not all the time and often not accurately, and the sources themselves are not infrequently suspect. The whole enterprise cries out for responsible editing and clarification of the credentials (literally, certification that a person is to be believed) of the contributors—which neither *Wikipedia*'s resources nor its operational philosophy allow.

And then there is blogging. In the wide open spaces of bloggery, a great many fresh (often *extremely* fresh) perspectives on practically every kind of issue are available if you know how to get to them, and getting to them is easy once you have acquired the knack of using a search engine. Moreover, bloggers with something to say can say it directly, and those who are good at it can obtain substantial audiences. Many of the organs of the traditional press now feature samples of what has been said in blogs about particular topics, although the task of surveying the range and locating consensus is much too huge for the means that newspapers and magazines are able to devote to it. The fact that they try, however, is an indication that many of their readers (and their editors) think that some kind of intelligent sorting and evaluation is necessary. In some ways, the evaluation is like book or film reviewing. In other respects, it is like the process of anthologizing: presenting excerpts from the approved blogs, in much the same way that innumerable online sources reproduce regular newspaper sources. (As we saw earlier, part of the current crisis of journalism revolves around the question of what will happen to internet news media if the newspapers on whose professional, expensive reporting they depend were to stop functioning.) In any event, there is a perceived need for guidance, summary, and evaluation in order to allow readers of blogs to make sense of what they are reading and responding to.

This is only one of the ways in which a structure like the classic public sphere seems to be necessary even in the internet world. There are others. The expression of a huge range of opinions on particular issues is certainly admirable from the standpoint of democracy, but action, change, the exertion of influence all depend on creating consensus, however partial. This is one of the functions of the core public sphere, one that—again—the mere presence of the internet does not make redundant. The internet has afforded the opportunity to participate actively in the conversations of the public sphere much more widely than ever before, but this makes the need for structures that formulate the options for public action that follow from the conversations and that subject them to informed, reasoned criticism even more pressing. Equally pressing is the need for a public space within which the claims of experts and conflicting interpretations of research can be discussed by people who have proven their good sense but are not necessarily experts. This has been the particular specialty of the core public sphere. For these and a host of other reasons, it appears to be the case that not only is open, informed public discourse necessary in a modern society, but so, too, is a construct that performs most of the functions of the core public sphere. The development of new technologies holds out the possibility (although not the guarantee) that those functions can be performed better and more democratically, but not that they can be done without.

Resolving the Crisis of the Public Sphere

It seems obvious that, given its less-than-stellar performance in the past decade or so, the core public sphere needs to be changed. The current financial difficulties of the quality press and competition from the internet media are, or could be, opportunities for improvement. For one thing, embracing the internet and other new technologies could make it possible for the discussions of the core public sphere to reach more people and for more people to take an active part in them.

But "active" is not enough. More people have to take an *intelligent* part in the conversation. The elitism that has characterized the classic core public sphere has been based, in both theory and practice, on the supposition that only a limited number of people have the ability, the background, and the interest to participate in intelligent public discourse, and that even among those who have these qualifications, some are (not to mince words) crackpots who need to be clearly identified as such. This elitism may be excessive, snobbish, misguided, even repellant, but in a fundamental sense it is not wrong. To take part in an intelligent conversation, you do actually have to know something, to be able to formulate your ideas in a coherent way that can be understood and therefore criticized by your interlocutors, to comprehend and adopt a common set of practices governing argumentation and response while maintaining an adequate analytical distance from the consensus that usually accompanies such practices. Not everyone meets all of these qualifications, whether because of lack of ability or because of defective education or temperament or because of absence of interest. But a vastly greater number of people do, or could, qualify than most of the builders of the nineteenth-century core public sphere believed—or than most of the people who interpret higher education as meritocratic sorting believe today. Thus, one goal for broadening the serious part of the public sphere would be to increase the pool of people who can participate in it intelligently. This is obviously a task for education (which we will discuss shortly), but it goes beyond that. It includes identifying and encouraging the adoption of appropriate models for discourse within the new media of the public sphere. This is unlikely to happen on a large scale simply as a result of millions of bloggers interacting with each other. It will require conscious action.

We should also deliberately try to use the internet to create the kind of community built around conversation that John Dewey insisted in the 1920s was the essence of the public.[2] There are certainly specific communities of bloggers, members of online (and offline) discussion groups, politically committed people, and so on, but there is no general community of people outside the traditional narrow core public sphere who engage in comprehensive discussions of significant matters with each other. How could there be, if all the models (including those imitated on the internet) are fragmentary—built around particular interests, points of view, ideologies, quirks? Until we seriously think about doing things in a new way rather than simply adapting what we already do for use with new technologies, we will not be able to respond to this need.

We must deal with the fact that many concepts and practices of public discourse that may in themselves be unobjectionable or even highly desirable nevertheless obscure the nature of the public sphere. This is, for example, the case with the cult of objectivity. Intelligent public discourse requires input from research conducted according to the procedures of objective evaluation employed by particular disciplines or professions or sciences. A significant part of the threat to public discourse posed by the incipient demise of major newspapers is that the burden of careful, professional journalistic research has been primarily borne by them, and nothing has offered itself as yet to take their place. We rightly fault news networks that egregiously violate existing standards of journalistic research (especially when they claim to be doing otherwise). We rightly object when scientific research sponsored by the federal government is "reinterpreted" for the public to correspond to the political or religious outlook of the people in power. At the same time, as we have seen, our desire to promote objectivity has tended to obscure the function of the core public sphere and to support attempts to segregate the functions of research from public discussion, so that those who do research are supposed to intervene in public discussion *only* as providers of scientific "truth." This kind of thinking has to be consciously combated.

The core American public sphere has never been isolated from the rest of the world, or at least from the Atlantic world. As we have seen, it started as a branch of an Atlantic public sphere in the nineteenth century and never lost its transatlantic connections. One of the factors leading to substantial gaps between the core public sphere and the broader discursive public in the United States has been the much greater insularity of the latter. In the contemporary world, the core public sphere needs to be globalized even more thoroughly than it has been in the past, so that its conversation is coherently and intelligently connected with those in other countries. This is one of the areas in which the long-standing link between higher education and the public sphere has been very significant in the past and where we should expect it to be even more important in the future. But more important still must be efforts to globalize the *entire* American discursive public, beyond the boundaries of the core. The internet, much more than television or the other communications technologies of the twentieth century, has made this possible, as has the adoption of English as the global second language. The assumption by the United States of its leading role in world politics has made it absolutely necessary. To some degree, the globalization of the American public sphere has already occurred, but it is hampered by, among other things, a measurable deficiency of knowledge about the rest of the world on the part of the broad American public. This frequently reported deficiency must be dealt with, obviously by education, but also by integrating the core public sphere (already highly globalized) more fully with the rest of the discursive public.

The last point just noted may the most important of all. As we have seen, the distinction between the core public sphere and the broad discursive public is a loose one; the quality media and other elements of the core are probably more accurately visualized as clustering toward one end of a continuum than as constituting an entirely separate category. Nevertheless, the distinction is real, and despite the fact that quite a lot of people move between the spheres and some inhabit both professionally, the majority of the participants in each tend to see the others through the medium of a set of not-very-positive stereotypes. To some extent, such stereotypes will always exist and some sort of distinction between the elite and the popular will always be imagined. But the distinction does not have to be seen as a wide gap. Another task that could be performed more effectively by public higher and secondary education, together with public broadcasting and other media, would be to continue to narrow the gap.

Change in Public Higher Education as a Means of Resolving Crisis

What has just been said has been expressed in very broad terms. In order actually to do something, to repair or replace or supplement the deficient elements of the American public sphere, generalities are obviously not sufficient. What I want to do now is to suggest specific ways in which *higher education* could contribute to a desirable resolution of the crisis of the core public sphere. To make an effective contribution, universities would have to change some of the ways in which they operate and some of the ways in which their faculty and leaders look at their place in the world. In compensation for the effort, higher education could go a considerable distance toward satisfactorily resolving its own crisis. Although much of what I have to say could apply to both public and private institutions, I will focus on public universities because they have been the focus throughout the book and because they are the institutions that would have to bear most of the burden of realizing what I suggest.

Emphasize the Vital Role of Public Universities in the American Public Sphere

As this book has demonstrated, higher education has been connected in both conception and practice to the core American public sphere ever since the latter was constructed in the second half of the nineteenth century. Its importance in this regard has grown substantially since then. *Public* higher education has been the principal means by which the core public sphere has been democratized, thereby fitting it better to perform its function in modern society. The accomplishments of higher education in expanding and supporting public discourse should be clearly articulated by both universities and the other components of the public sphere.

I am not suggesting that public universities should stop explaining how important they are to the economy. Universities, public and private, are vital to the economic life of the country and the rest of the world. But as we have seen, that is not their only or their main significance, and focusing single-mindedly on their economic function has seriously threatened the performance of their other roles and distorted the conceptions that frame the ways in which public universities are financially supported. I am not suggesting that public universities should stop advertising themselves as centers of research, whether with economic or noneconomic implications. Much of what is valuable in the modern world has been produced by university research. I do think that state-supported universities should examine very critically the extent to which research has come to be linked—conceptually as well as practically—to their economic and revenue functions and entertain the possibility that the costs of emphasizing the linkage may outweigh the advantages. But there is no reason that public universities should stop engaging in research or citing it when they explain their importance. I am not suggesting that public universities should stop making the argument that they are vital to social progress in the United States. If anything, they have been too quiet about that in the past ten years.

What I *do* want to suggest on the basis of the preceding chapters is that public universities acknowledge to themselves and announce emphatically to the world at large that their relationship specifically to the core public sphere (not just to the broad "public" in its many senses) constitutes a function that is at least as important as any other. If they could get this point across outside their institutional walls, and if in doing so they stimulated a larger effort to publicize the importance of the core public sphere and the dangers that the latter faces, they should be able to make a much stronger case for comprehensive state support than they do now. They might also diffuse some of the pressure that arises from being considered mainly as engines of economic development. Their ability to do this convincingly, however, would depend heavily on the willingness of faculty and administrators to make changes in the modes in which they think about higher education, and also on their ability to act on those changes.

Acknowledge Participation in the Public Sphere as a Significant Aspect of an Academic Career

A substantial segment of the active personnel of the core public sphere in the United States, as elsewhere, actually consists of faculty at universities and colleges. In some cases, a professor's prominence as a public intellectual is recognized and rewarded by his or her institution, and in other cases universities deliberately recruit people who have not followed standard academic careers but who have distinguished themselves by contributing to the public conversation. For the most part, however, writing for quality and opinion magazines, publishing "popular" as opposed to scholarly books, spending time discussing issues rather than, say, working in the laboratory all have generally been treated as peripheral to the principal tasks of faculty, as (at best) forms of "public service" not categorically distinguishable from speaking at the monthly meeting of the local Lions club or acting as a judge at a high school science fair. Not that there is anything wrong with Lions clubs or science fairs, but the fact that engagement in the public sphere has regularly been classified in the same way suggests the lack of importance accorded to the latter in academic ideology and in the practices of universities. Professors are expected to perform "service," but the service that generally counts the most toward tenure, promotion, and pay raises is membership on university committees. Service in its other aspects is usually considered optional and dispensable. Especially in universities that are not confident about their academic status or whose administrators worry about how legislators will react to any hint of political activity by faculty, expressing opinions outside of one's professional field (or even within the field, if a partisan slant can be put on what is said) is often disparaged or tacitly (sometimes not so tacitly) discouraged. All of this needs to be rethought and, as much as possible, changed: the policies, the attitudes that underlie them, and the elements of academic ideology in which the attitudes are embedded.

I want to be clear that, in saying this, I am not advocating a wholesale rejection of "pure" research in favor of applied research. That is a different matter, involving a conflict of views within the same conceptual framework: whether greater value should be placed on research done "for its own sake," which may or may not have practical relevance for technology or policy, or on research designed from the start to solve specific technological, social, or policy problems. Either way, the central activity of the researcher is envisioned as creating a product that is presented to the world as a consequence of the objective endeavors of the researcher within the university. As we have seen, this is not the same thing as taking part in public debate as an intelligent, educated person who may or may not possess a professional qualification in the field under discussion. Research is an important function of the university and obviously a significant resource for the public sphere, but what is needed in addition is an emphasis on the other, more direct aspect of participation and its acceptance as a central and legitimate part of a faculty member's work.

This would not require a lowering of standards for assessing faculty performance in any meaningful way. Faculty engagement in public discourse should be evaluated according to recognized criteria of quality. Mere expressions of opinion on issues should not be sufficient to rate highly. At present, because the service component of a university instructor's workload is generally considered to be peripheral, reviews of service tend to be fairly perfunctory—often little more than a check-off or an assessment of amount of time put in. It would not, however, be difficult to rate the quality as well as the quantity of contributions to public discussion in the same way that scholarly publications are weighed, especially if such contributions were considered in a category of their own. Care would have to be taken that political disagreement did not affect the result, but again, pretty much the same mechanisms for doing this that are employed in evaluating standard research could be used for participation in the public sphere. Most universities award prizes to faculty for especially meritorious performance in research, teaching, and service; they could do the same thing for taking part in public debate.

Universities should strongly emphasize the fact that one of the principal reasons for the system of tenure is to protect the expression of opinions on issues, and not just on scholarly or academic matters. Not only should they take the quality of such expressions into account and assess them rigorously in deciding whether or not to award tenure, but they should make it clear that possessing tenure creates something approaching an obligation on the part of a professor to make use of the protection that tenure affords.

Emphasize the Public Sphere in the Undergraduate Curriculum

To make their commitment to public discourse credible, to contribute to general awareness of the importance of the core public sphere, and to help increase the latter's effectiveness, universities should change some of the ways in which they undertake the teaching of undergraduates. Alterations are needed in general education programs and in major concentrations, and also in the (usually optional) area of the curriculum labeled "service learning." The key aims in each case should be to bring consciousness of the public sphere to center stage and to encourage students to develop the skills and acquire the knowledge needed to participate actively in public discourse. These were among the goals that the university reformers of the nineteenth century sought to achieve in restructuring university education. They were not entirely lost from view in the twentieth century, but they tended to be subsumed under other categories, especially education in citizenship, social science, and expository writing. They should be brought out in the open and named as such, and the means by which they are achieved should be consciously laid out.

The idea that general education should promote "citizenship" has a complicated history. Some versions of it have focused on inculcating appropriate political attitudes and beliefs in students, together with basic knowledge about the ways in which the political system is supposed to operate. Traditionally, required courses in government and sometimes in American history have been seen as the means of accomplishing such tasks. This version of teaching for citizenship is probably misguided. Basic knowledge is (or should be) available in high school. Moreover, even if one agreed that citizenship training should be indoctrination (which I do not), by the time most students have attained college age they tend to resist indoctrination anyway. At the university level, education in citizenship more appropriately aims at teaching students the critical skills necessary to evaluate intelligently both the beliefs that they have been taught and the issues that their society currently faces, and also at affording knowledge needed to make the evaluation useful. That is what most required courses in American history and government at public universities actually attempt to do, regardless of what the people who devised the requirement (often legislators) may have intended. I think it is unfortunate that, perhaps because of worries about the intention to indoctrinate that may lie behind a "citizenship" requirement, these really useful elements are absent from the mandatory curriculum in many institutions (at my own university, for instance).

But beyond these matters, another significant question needs to be asked: what actually *is* "citizenship"—at any rate, the citizenship that is meant when we say that a college education should develop good citizens? It is sometimes described as knowing what people need to know in order to be able to vote. Sometimes it is seen as a pattern of virtuous political behavior. (Don't cheat at elections, don't riot afterwards, accept properly elected leaders even if you didn't vote for them, and so forth.) But voting and behaving properly constitute only part of what thinking, critical participants in political life do. Mostly, they take part in the conversations of the public sphere. Voting, and even deciding how to vote, takes relatively little time and often not a great deal of mental effort. Attending to debates on policy options and exploring issues take much more time and effort and a higher level of knowledge and intellectual skill. At the university level, education for citizenship is primarily education for the public sphere; we would do well to acknowledge that fact openly and explicitly organize part of the curriculum around it. Education for the public sphere requires critical analysis, but it must also contain knowledge of a broad range of subject matter: the practices and institutions of politics, yes, but also of many other topics. In my experience, organizing such topics around historical study is a very effective way of structuring this kind of instruction, but there are others.

One of them that has been favored for many years is education in the social sciences. Julie Reuben has described how, between the late nineteenth century and the second quarter of the twentieth century, social science came to replace moral philosophy in the American undergraduate curriculum—as to a considerable extent it did in the intellectual life of universities in general.[3] More recently, something similar has happened with regard to education for citizenship. Responding both to the view that training for citizenship is essentially indoctrination and, in many cases, to a widespread belief that all university instruction should be "objective," many universities have in effect substituted requirements for courses in social sciences (or sometimes for interdisciplinary courses on social problems that draw heavily on the social sciences) for ones that deal more specifically with preparation for citizenship.

Knowing about the approaches and assumptions of social sciences is, obviously, very useful for understanding contemporary public discourse, but it is not a substitute for explicit preparation for engaging in the conversations of the public sphere—unless the social science courses offered for general education address the public sphere directly. This can be done, and frequently is done in courses that focus, for example, on using social science to evaluate current national debates.[4] Too often, however, the courses recommended or required for general education are the survey courses in particular disciplines, the point of which is typically to lay out the elementary theory of the discipline in question. This is unobjectionable in itself, but it has more to do with understanding a mode of academic thinking than with active engagement in public discourse. The difference between the two kinds of aim ought to be explicitly recognized, and both should be incorporated in the curriculum.

Courses on writing are an integral part of practically all general education programs in American colleges and universities. When they began in the nineteenth century as an outgrowth of traditional training in rhetoric, their function as adjuncts to the public sphere was straightforward and well understood: they taught people how to express themselves in print on matters of public significance. In the course of the twentieth century, the function was not so much lost as it was shouldered aside, as writing instruction became a task of English departments that defined themselves largely in terms of the study of literature, criticism, and the humanities. Most general education writing programs today devote some attention to analyzing newspaper editorials and preparing essays and research papers on contemporary issues, but only rarely do these activities constitute the central thrust of the endeavor. More emphasis on writing and acting in the public sphere would, I think, be useful to students and to the public alike. It would also, once again, manifest the seriousness of a university's commitment to supporting public discourse in a fundamental way. Actual attempts to do this—to orient expository writing around the analysis of significant issues—have encountered substantial resistance, from both faculty who think that the primary focus should be on academic writing and a wide range of critics who object to "politicizing" the curriculum.[5] This resistance needs to be overcome.

Emphasizing the public sphere should not be limited to undergraduate general education; it should extend also to undergraduate major and graduate programs. This does not necessarily mean that new degrees specifically oriented toward public discourse should be created. There would be nothing wrong with that, but universities are already well equipped with concentrations in journalism, communication, and media studies. Rather, some of the more traditional departments should consider taking specific account of the public sphere in their courses and degree programs. My own department, for example, has recently instituted a concentration in "public history" in its Master's degree curriculum that is oriented, at least in part, toward the public sphere. Many colleges and universities in the United States have also emphasized "service learning" experiences for undergraduates as integral parts of the curriculum. It would be useful if such programs could provide experiences in the public sphere. Internships on newspapers, for example, are usually represented as preprofessional training for prospective journalists, not as "service." There is no reason that they could not be redefined as a form of service to the public—especially if universities were to follow the next set of suggestions.

Participate Directly in the Public Sphere

As we have seen, the new-model American universities of the late nineteenth century intentionally supported the core public sphere. Their personnel (including some of their presidents) regularly contributed articles to the leading organs of public discourse. In practice, not only have university faculty (not, for the most part, including their presidents) continued to be major contributors, but higher education in general has become a central element of the core public sphere. As we have also seen, a host of factors has prevented this centrality from being fully recognized or made a significant part of what universities *consciously* do as institutions. Some of these factors (dislike of being regarded as partisan or offending a range of ideological sensibilities, adherence to the idea that universities should only be locations where objective science is practiced, fear of offending legislators and voters) will never be completely eliminated, but if it is important to maintain an effective public sphere, something needs to be done to counteract their effects.

Universities as institutions should forcefully assert their role in public discourse, not only with words but with actions, and they should work to change public assumptions about what that role is. They should be much more aggressive than they were between 2001 and 2005 about sponsoring informed discussions of national policy—especially foreign policy. They should be willing to sponsor and publicize discussions of topics that most institutions of the American public sphere tend to avoid and to entertain a wider range of opinions—with suitable opportunities for critical response—than those institutions usually allow.[6] This would require universities and their faculties to remove, or at least to make more elastic, the boundaries between what is and what is not acceptable for the officially recognized conversations that go on within their institutions. Doing so will not be easy, but it is necessary.

Universities should also greatly expand the range of ways in which they support the publication of ideas and positions on issues outside the conventional confines of academic research. In recent years, university presses appear to have become much more engaged in current public discourse than they once were, publishing books that take positions on matters under political dispute while attempting to apply the same standards that they use for purely academic books. This is entirely appropriate. Other elements of universities could do a great deal more along the same lines. Universities could move into online publication much more extensively and on a broader array of topics than most of them have done so far. They could, for instance, offer to intervene in the current newspaper crisis.

If high-quality daily newspapers that undertake journalistic research are a public good for which there is no known substitute, and if such newspapers cannot remain solvent and effective as private enterprises in the market, would it not make sense for some of them to develop partnerships with universities in the same cities or regions? It has already been proposed that the government should subsidize newspapers directly, although the images that the idea raises of the controlled state press in nondemocratic countries make realizing the proposal very doubtful.[7] But public universities are already subsidized by the state, and they possess not-insignificant capacities to withstand political efforts to interfere in what they do. (They could use more, of course, which would be necessary in any case if higher education were to emphasize its role in the public sphere as a central part of its formal mission.) One could imagine arrangements whereby some newspapers became essentially institutes attached to universities, with funding from subsidies that were part of the university's state appropriation as well as from grants, private sources, subscriptions, and some advertising.

Interaction between newspaper-as-institute and the rest of the university might go considerably beyond financial and administrative connections. The most vital and one of the most expensive functions of newspapers—in-depth reporting—is essentially research, which is something that university faculty, students, and institutes also do. Indeed, newspapers rely very heavily on university research for background to many of their articles and on university faculty for expert commentary. Collaboration between academic and journalistic research, possibly in some instances even integration, might be an advantage for both as well as for the public. Close partnership could also benefit university instruction, encouraging the kind of emphasis on education for the public sphere suggested earlier. Reporters and staff writers (many of whom have recently lost their jobs) could, for instance, teach writing to undergraduates as part of their full-time responsibilities. Many other avenues of cooperation could be envisioned. This would be especially true if universities (and newspapers) adopted the following suggestion.

Move Vigorously into Cyberspace

Higher education has been going in this direction for some time, although mainly by making relatively minor adaptations in the ways in which universities normally do things in order to take advantage of possibilities afforded by the internet: courses taught online, but still in traditional formats; journals published online, mainly to save costs and in more or less standard forms; online conferences; and so forth. Some universities have been bolder, rethinking their modes of instructing and of eliciting understanding, as, for example, by adopting "competencies" as categories of learning and evaluation and abandoning the notion that higher education in the United States must take place within the confines of the semester or quarter course granting three or six hours of credit. A great deal more could be done, especially from the standpoint of renovating and further democratizing the public sphere.

Public universities could, for example, seriously consider putting a much larger proportion of their academic programs into formats that encourage participation by people outside their institutional walls, either as students or in other capacities. The internet provides the most obvious medium for such participation, but it is not the only one. "Hybrid" programs involving a combination of individual study, online communication with instructors and other students, and occasional face-to-face meetings are already in operation in some places and could be expanded. (There is no threat in this to the residential campus. The appeal of a "standard" college education and the prestige that accrues from having had one will ensure that it does not disappear.) Programs involving discussion of contemporary public issues, whether structured as parts of degree curricula or otherwise, could help to break down the walls between the core public sphere and the broader discursive public. Universities could also move more fully into online publishing. They could support an expansion of electronic periodicals devoted to topics significant to the public sphere, encourage university presses to move in a similar direction, and make sure that their libraries assign a high priority to subscribing to online publications.

Universities could perform a real service to the public sphere by taking upon themselves the task of solving some of the problems of credibility and standards in the discourse of the cyber world that were noted earlier. Acting individually or collectively, they could organize systems for reviewing websites and blogs, encourage their faculty and students to create and contribute to sites established under university (or university and newspaper) sponsorship, and establish extensive partnerships with entities like *Wikipedia*. Steps have already been taken in these areas, but they are only a beginning of what could become a really significant, and entirely appropriate, connection between higher education and the public sphere.

The suggestions that have just been given are only a sample of what might come out of a really serious national discussion of the ways in which higher education could usefully develop its role in public discourse in the age of the internet. Others topics of a more general nature present themselves as questions for further discussion. For example, how could American higher education build on its long-standing function as a medium of Atlantic (and now global) intellectual exchange so as to reduce the isolationism that still characterizes so much of political and cultural conversation in the United States? Because the boundary of isolationism has tended to follow the zone of demarcation between the core and the broad public spheres, this question is related to another: how could higher education contribute in general to breaking down some of the distinctions between the two spheres in such a way that nothing of value is lost from the traditional core? There is also a paradoxical question: how could universities use their central role in the core American public sphere to try, perhaps in the long run, to *reduce* the dependence of public discourse on academia and higher education? It would be far better for the United States if, as in some other countries, more of the public conversation were carried on outside the academic world. Considering the current distribution of tasks in the core public sphere, however, it is clear that a substantial part of the thinking about this matter must take place in universities. And finally, how could universities contribute to building the national and global discursive communities that John Dewey—rightly, I think—saw as essential to an effective public in the modern world? If universities, especially but not exclusively public universities, can contribute to answering these questions, if they can change their ways of doing things so as to acknowledge and better perform their roles in the public sphere, they can go a long way toward resolving favorably both the general crisis of the discursive public and the more specific crisis of public higher education.

Notes

Introduction

1. See Alasdair Roberts, *Blacked Out: Government Secrecy in the Information Age* (Cambridge: Cambridge University Press, 2006) and Albert O. Hirschman, *Shifting Involvements: Private Interest and Public Action* (Princeton: Princeton University Press, 2002). See also Clint Hendler, "How to Keep Secrets: Obama Tries to Get Classification Right," *Columbia Journalism Review* website, September 2, 2009, www.crj.org/campaign_desk/how_to_keep_secrets.php?page=all (accessed January 1, 2010), and Lee White, "Obama Administration Issues Sweeping Open Government Directive," *Perspectives on History* 48, 1 (January 2010): 18–19.
2. Al Gore, *The Assault on Reason* (New York: Penguin, 2007). See also Herbert J. Schiller, *Culture, Inc.: The Corporate Takeover of Public Expression* (New York: Oxford University Press, 1991).
3. Katherine Q. Seelye, "Drop in Ad Revenue Raises Tough Question for Newspaper," *New York Times*, March 26, 2007; John Morton, "Facing the Future: Newspapers Are Making Necessary Changes to Endure in the Internet Era," *American Journalism Review* 29, 2 (April 2007): 68.

Chapter 1

1. The book in which Habermas introduced the concept of the public sphere is Jürgen Habermas, *Strukturwandel der Öffentlichkeit* (Neuwied: Luchterhand, 1962), translated as *The Structural Transformation of the Public Sphere: An Inquiry into a Category of Bourgeois Society* (Cambridge, MA: MIT Press, 1989). Citations here refer to the English-language edition. One of the reasons that the English term "public sphere" has caught on internationally is its implication of spatiality. It can be thought of as a set of concrete places, with physical attributes and boundaries and specific historical referents, rather than as an abstraction, which *Öffentlichkeit* suggests. Habermas clearly intended this kind of specificity.
2. Habermas, *Structural Transformation,* pp. 37, 54.
3. Ibid., pp. 27–43.
4. Ibid., pp. 141–235.
5. See, for example, several of the contributions to Craig Calhoun, ed., *Habermas and the Public Sphere* (Cambridge, MA: MIT Press, 1992), especially those of

Michael Schudson (pp. 143–63), Mary P. Ryan (pp. 259–88), and Geoff Eley (pp. 289–339).
6. For an idea of the way in which the core-periphery distinction and the notion of diffusion were viewed in the late nineteenth century, see "The Newspaper and the Organ," *Century* 29, 3 (January 1885): 461–62. An "organ" is defined in the article as an intellectually respectable, quality journal (like the *Century*, presumably) that attempted to be nonpartisan with regard to the political parties, although not neutral with regard to issues.
7. This matter will be discussed in chapters 5 and 6. See Thomas Bender, *Intellect and Public Life: Essays on the Social History of Academic Intellectuals in the United States* (Baltimore: Johns Hopkins University Press, 1993), pp. 49–77.
8. For current examples, look at any issue of any major opinion journal. Here is one: Niall Ferguson, "Sinking Globalization," *Foreign Affairs*, March/April 2005. Available at www.foreignaffairs.com/articles/60622.
9. On professionalization in America, see Burton Bledstein, *The Culture of Professionalism* (New York: Norton, 1976), and Thomas Haskell, ed., *The Authority of Experts* (Bloomington: Indiana University Press, 1984).
10. On professionalization in the context of universities (not just in America), see Konrad H. Jarausch, *The Transformation of Higher Learning, 1860–1930: Expansion, Diversification, Social Opening, and Professionalization in England, Germany, Russia and the United States* (Chicago: University of Chicago Press, 1983).
11. Charles Taylor, *Modern Social Imaginaries* (Durham, NC, and London: Duke University Press, 2004), pp. 83–99, gives an interpretation of the relationship of the public sphere to modernity.
12. This arrangement was described in the 1880s by the British observer James Bryce, who believed that it had, in fact, been satisfactorily established. See James Bryce, *The American Commonwealth* (3 vols.; London and New York: Macmillan, 1888), 3, pp. 3–63.
13. See Curtiss S. Johnson, *Politics and a Belly-full. The Journalistic Career of William Cullen Bryant, Civil War Editor of the New York Evening Post* (New York: Vantage Press, 1962).
14. One change that *has* occurred is a growing awareness that there is something wrong with public discourse in the United States, that part of the problem lies in the structure of public discussion, and that an equally important part is the apparent unwillingness or inability of members of the public to think carefully and critically. These points are made cogently in Al Gore's *The Assault on Reason.*
15. John William Tebbel and Mary Ellen Zuckerman, *The Magazine in America, 1741–1990* (New York: Oxford University Press, 1991), pp. 319–22.
16. Ibid, pp. 318–19; www.theatlantic.com/about/atlhistf.htm (accessed April 28, 2009).
17. Journalism.org, The State of the News Media: An Annual Report on American Journalism, 2008: www.stateofthenewsmedia.org/2008/printable_magazines_opinions.htm (accessed January 5, 2009).
18. Tebbel and Zuckerman, *Magazine in America*; David Seideman, *The New Republic: A Voice of Modern Liberalism* (New York: Praeger, 1986).

19. Rafat Ali, "Earnings: Salon.com's Revenues Decrease; Losses Shrink," paidContent.org, August 10, 2005, www.paidcontent.org (accessed April 28, 2009).
20. Megan Woolhouse, "*Globe, Herald* Circulation Tumbles," *Boston Globe,* October 27, 2009, citing the semiannual report of the Audit Bureau of Circulations. The daily editions of the *Boston Globe* and the *Boston Herald* fell 18.5 percent and 17.5 percent, respectively, during the reporting period.
21. A well-reasoned argument for subsidy appears in John Nichols and Robert W. McChesney, "The Death and Life of Great American Newspapers," *Nation,* April 6, 2009: 11–20, which also gives a cogent analysis of the current problems of the newspapers.
22. See, for example, Herbert J. Schiller, *Culture, Inc.: The Corporate Takeover of Public Expression* (New York: Oxford University Press, 1991).
23. This was, for example, one of the themes of the famous "Hutchins Report" on the media, issued in 1947. See Stephen Bates, *Realigning Journalism with Democracy: The Hutchins Commission, Its Times, and Ours* (Washington, D.C.: Annenberg Washington Program in Communication Studies, 1995).
24. John Morton, "Facing the Future: Newspapers are Making Necessary Changes to Endure in the Internet Era," *American Journalism Review* 29, 2 (April 2007).
25. Al Gore, *Assault on Reason* (New York: Penguin, 2007), pp. 245–73.
26. Kenneth Li and Andrew Edgecliffe-Johnson, "Murdoch Vows to Charge for All Online Content," *Financial Times,* August 6, 2009; Lymari Morales, "Cable, Internet News Sources Growing in Popularity," *Gallup,* December 15, 2008 (accessed at http://www.gallup.com/poll/113314/).
27. Morales, "Cable, Internet News Sources."
28. Gore, *Assault on Reason,* pp. 16, 20, 34–36, 76, 97, 102, 246.
29. See Steve M. Barkin, *American Television News: The Media Marketplace and the Public Interest* (Armonk, NY: M.E. Sharpe, 2002).
30. For a barebones summary of the workings and financing of American public radio and television, see L.R. Ickes, ed., *Public Broadcasting in America* (Hauppauge, NY: Nova Science Publishers, 2005).
31. See Jerold M. Starr, *Air Wars: The Fight to Reclaim Public Broadcasting* (Philadelphia: Temple University Press, 2001).
32. Jerome De Groot, *Consuming History* (New York: Routledge, 2008), p. 185.
33. See Bender, *New York Intellect,* pp. 296–302, and Robert Westbrook, *John Dewey and American Democracy* (Ithaca: Cornell University Press, 1991), pp. 195–227 for the First World War. See also Edwin R. Bayley, *Joe McCarthy and the Press* (Madison: University of Wisconsin Press, 1981), and Clarence R. Wyatt, *Paper Soldiers: The American Press and the Vietnam War* (Chicago: University of Chicago Press, 1995).
34. With regard to 2005–6, the *Nation* had opposed the Iraq War and Bush from the start, the *New Republic* supported both; the *Washington Post* was strongly on the side of the administration, the *New York Times* maintained a position of neutrality. As far as I could tell at the time (and I have had no reason since then to think that I missed anything), the shift was triggered by two factors: public perception of the administration's incompetence in responding to Hurricane Katrina, and

the shaming of the administration by Cindy Sheehan and other relatives of soldiers killed in Iraq. Neither of these developments owed much—if anything—to the central institutions of the core public sphere. The events were reported in the general press and on regular network television, and the public's reaction to them was not led by any publication in particular.

35. Ronald Steel, *Walter Lippmann and the American Century* (Boston: Atlantic Monthly-Little, Brown, 1980), pp. 557–84.
36. Gore, *Assault on Reason*; Schiller, *Culture, Inc.*; Alasdair Roberts, *Blacked Out: Government Secrecy in the Information Age* (Cambridge: Cambridge University Press, 2006).
37. These suggestions have been around a long time. They were discussed, although not definitively recommended, in the Hutchins Report. Commission on Freedom of the Press, *A Free and Responsible Press: A General Report on Mass Communications—Newspapers, Radio, Motion Pictures, Magazines and Books* (Chicago: University of Chicago Press, 1947).
38. People have tried to do this. In 1996, the *National Review* ran a remarkable series of articles that offered various arguments for standing down in the war on drugs and treating drug addiction in a different way. The series seems to have had little effect. See William F. Buckley Jr. et al., "The War on Drugs is Lost," *National Review*, February 12, 1996.
39. Walter Lippmann, for example, made sophisticated versions of this argument in the 1920s. See Walter Lippmann, *Public Opinion* (New York: Free Press, 1965; original edition 1922), pp. 272–75, and Walter Lippmann, *The Phantom Public* (New York: Harcourt, Brace, 1925), pp. 13–53.
40. Every year, the *Nation* sponsors a cruise during which readers—or anyone else—can spend several days with the magazine's editors, many of its contributors, and some of its "heroes." See www.nationcruise.com. This seems to me to symbolize the tendency suggested here: discursive communities on separate boats, enjoying conversations among themselves but not seriously engaging the people on the other boats—and even more important, inviting the public to embark on the boats and take part in the conversations in the discursive communities' terms, rather than disembarking and trying to speak the languages used by most of the public.

Chapter 2

1. Throughout this book I will generally use the term "public university" to refer to all types of state-supported, degree-granting colleges and universities with four-year undergraduate programs. I do this entirely for convenience, because "public colleges and universities" is awkward when repeated. When the discussion extends to two-year junior or community colleges, I will indicate this specifically.
2. Jeffrey J. Selingo, "State Budgets Are Likely to Squeeze 2-Year Colleges," *Chronicle of Higher Education* 55, 11 (November 7, 2008); "13 Reasons Colleges Are in This Mess," *Chronicle of Higher Education* 55, 27 (March 13, 2009).
3. Jean Evangelauf, "State Appropriations for Higher Education See Biggest Jump in Decades," *Chronicle of Higher Education* 54, 19 (January 18, 2008).

4. "State Higher Education Appropriations Reflect Recessions," *National Association of College and University Business Officers*, November 4, 2006, accessed at www.nacubo.org/Initiatives/News/State_Higher_Education_Appropriations_Reflect_Recessions.html.
5. Beth Healy, "Student grants get a boost," *Boston Globe*, March 24, 2010.
6. Ibid.; Michael D. Shear and Daniel de Vise, "Obama Announces Community College Plan," *Washington Post*, July 15, 2009.
7. See Liza Featherstone, "Out of Reach: Is College Only for the Rich?" *Nation*, June 29, 2009, pp. 11–14. See also the editorial in the same issue, p. 3.
8. This is shown convincingly in Christopher Newfield, *Unmaking the Public University: The Forty-year Assault on the Middle Class* (Cambridge, MA: Harvard University Press, 2008), pp. 208–19.
9. I took notes on the back of the program.
10. See Derek Bok, *Universities in the Marketplace: The Commercialization of Higher Education* (Princeton: Princeton University Press, 2004), and Richard H. Hersh and John Merrow, eds., *Declining by Degrees: Higher Education at Risk* (New York: Palgrave Macmillan, 2006).
11. Newfield, *Unmaking the Public University*, pp. 208–19.
12. For example: "Value of a Liberal Arts Education," College of Arts and Sciences Advising Homepage, University of Hawai'i at Manoa, www.advising.hawaii.edu.edu/artsci (accessed August 20, 2009).
13. See the welcoming statement by the president of the University of Massachusetts on the university's website: www.massachusetts.edu/po (accessed August 20, 2009).
14. "Regents determine CEO salaries," news release, June 26, 2003, on the website of the Kansas Board of Regents: www.kansasregents.org/news/2003/ceo.html (accessed August 20, 2009).
15. Newfield, *Unmaking the Public University*, pp. 125–41.
16. See Albert H. Soloway, *Failed Grade: The Corporatization and Decline of Higher Education* (Salt Lake City: American University and Colleges Press, 2006); Joseph E. Stiglitz, *Globalization and Its Discontents* (New York: W.W. Norton, 2003).
17. See Robert Birnbaum, *Management Fads in Higher Education: Where They Come From, What They Do, Why They Fail* (San Francisco: Jossey-Bass, 2000).
18. Ibid., pp. 91–122.
19. James J. Duderstadt and Farris W. Womack, *The Future of the Public University in America* (Baltimore: Johns Hopkins University Press, 2003), p. 27.
20. See, for example, "Are state appropriations the problem?" Texas Public Policy Foundation, http://www.texashighered.com/state-appropriations (accessed August 20, 2009).
21. Malcolm W. Browne, "Supply Exceeds Demand for Ph.D.'s in Many Science Fields," *New York Times*, July 4, 1995.
22. Eric Hobsbawm and Terence Ranger, eds., *The Invention of Tradition* (Cambridge: Cambridge University Press, 1983).
23. This is beginning to change, especially among faculty in states such as California where the old arguments are so clearly inoperable in the budgetary process that faculty spokespeople have had to start refuting their opponents in detail. This can

be seen in Newfield's remarkable *Unmaking the Public University*, which dissects with great accuracy the positions of those who have attacked the University of California, affirmative action, and public higher education in general. Toward the end of this chapter, I will disagree with one of the arguments Newfield makes, but that should not be understood to reduce my admiration for his book, which shows the kind of action that public higher education should be taking to defend itself.

24. Salary increases and, in the case of associate professors, promotion to full professor can be withheld, but the possibility of losing one's job—the prime motivator in most occupations for meeting the full range of professional requirements—is to all intents and purposes ruled out.
25. Abraham Flexner, *Universities. American English German*, with an introduction by Clark Kerr (Oxford: Oxford University Press, 1968), p. 10.
26. This seems to have been the position, for example, of James Bryant Conant, president of Harvard from 1933 to 1953. Conant apparently saw his position as a necessary response to the temporary conditions of the Cold War and a way of countering McCarthyism rather than as a permanent policy. He was also willing to modify it in individual cases. See Wilson Smith and Thomas Bender, *American Higher Education Transformed, 1940–2005* (Baltimore: Johns Hopkins University Press, 2008), p. 458, and David Glenn, "John Kenneth Galbraith is Remembered as a Mentor and Scholar," *Chronicle of Higher Education,* May 1, 2006.
27. Newfield, *Unmaking the Public University,* pp. 51–106.
28. James O. Freedman, *Liberal Education and the Public Interest* (Iowa City: University of Iowa Press, 2003), attempts to do this, but not, it seems to me, as convincingly as it could—largely, I think, because Freedman is trying to defend the academic ideology on its own terms.
29. This is one of the central points made in Newfield, *Unmaking the Public University,* pp. 19–47.
30. For example, Flexner, *Universities,* pp. 52–68, and Robert Maynard Hutchins, *The Higher Learning in America* (New Haven and London: Yale University Press, 1936; reissued 1962 with a new introduction), p. 31.
31. On the origins and development of the meritocratic model, see Nicholas Lemann, *The Big Test: The Secret History of the American Meritocracy* (New York: Farrar, Straus and Giroux, 2000).
32. See Howard Ball, *The Bakke Case: Race, Education, and Affirmative Action* (Lawrence, KS: University of Kansas Press, 2000), and Patricia Marin and Catherine L. Horn, eds., *Realizing Bakke's Legacy: Affirmative Action, Equal Opportunity, and Access to Higher Education* (Miami: Stylus Press, 2008).
33. By unstated implication, the system will also preserve the existing social hierarchy, since the graduates of the less-prestigious universities will occupy the lower ranks of their professions and earn less money. The opportunity to rise in a profession is always open, but it must be assumed to be available only to a few of the graduates of the less desirable institutions. Otherwise, the logic that identifies other institutions as more desirable would be obviated: if they did not select the students who are intrinsically the best and if they did not give them the best preparation for

advancement, on what legitimate basis could they rest their claim to being most desirable? And if the best students did not advance in life higher than the others, then the claims of the professions themselves to maintaining standards would be suspect and the utility of meritocratic education would be undermined. Why go to Harvard if graduates of Memphis State do just as well professionally? (Or perhaps more to the point, why go to the University of Tennessee if graduates of Memphis State do just as well?).
34. Newfield, *Unmaking the Public University*, pp. 19–30, 51–67.
35. Ibid., pp. 80–106.
36. On the other hand, if they followed the advice of Stanley Fish, they would just hold to their traditional values and stick to what Fish regards as their task: teaching students and helping them to acquire critical and research skills. Unfortunately, universities can never stick only to those tasks, as we will see. As entities dependent on external support, as economic resources, and as the employers of faculty who will, like Fish himself, take part in public discourse, they cannot separate themselves from what goes on in the rest of society and cannot avoid having to justify themselves. See Stanley Fish, *Save the World on Your Own Time* (New York: Oxford University Press, 2008).
37. Newfield, *Unmaking the Public University*, pp. 265–75.

Chapter 3

1. A site clearly described as being outside academia. The *Spectator*'s famous intention was to "bring Philosophy out of Closets and Libraries, Schools and Colleges, to dwell in Clubs and Assemblies, at Tea-tables and in Coffee-houses." *Spectator*, No. 10, March 12, 1711, in Joseph Addison and Richard Steele, *The Spectator, Volumes 1, 2 and 3*, Gutenberg Edition, 2004, Ebook #12030, p. 72. On the *Spectator* and its relationship to public discourse, see Michael G. Ketcham, *Transparent Designs: Reading, Performance and Form in the Spectator Papers* (Athens: University of Georgia Press, 1985).
2. Jürgen Habermas, *The Structural Transformation of the Public Sphere: An Inquiry into a Category of Bourgeois Society* (Cambridge, MA: MIT Press, 1989), pp. 32–33.
3. Keith Michael Baker, "Defining the Public Sphere in Eighteenth-Century France," in Calhoun, ed., *Habermas and the Public Sphere* (Cambridge, MA: MIT Press, 1992), pp. 181–211.
4. Geoff Eley, "Nations, Publics, and Political Cultures: Placing Habermas in the Nineteenth Century," in Calhoun, ed., *Habermas and the Public Sphere*, pp. 288–339.
5. See John Breuilly, ed., *The State of Germany: The National Idea in the Making, Unmaking, and Remaking of a Modern Nation-state* (London: Longman, 1992), and Derek Beales and Eugenio Biagini, *The Risorgimento and the Unification of Italy*, 2nd ed. (Harlow: Longman, 2002).
6. Marshall S. Shatz and Judith E. Zimmerman, translators and editors, *Signposts: A Collection of Articles on the Russian Intelligentsia* (Irvine, CA: C. Schlacks, Jr., 1986).

7. Habermas, *Structural Transformation*, pp. 175–234.
8. John Clive, *Macaulay: The Shaping of the Historian* (New York: Knopf, 1973); A. L. Kennedy, *Salisbury 1830–1903. Portrait of a Statesman* (London: John Murray, 1953), pp. 30–63; Robert Blake, *Disraeli* (London: Trafalgar Square Publishing, 1998).
9. Raymond Huard, *La Naissance du parti politique en France* (Paris: Presses de la Fondation nationale des sciences politiques, 1996).
10. For a succinct nineteenth-century American description of the Carlton and Reform clubs (the headquarters of the British Conservative and Liberal parties, respectively), see "Leading London Political Clubs," *New York Times*, June 16, 1878, p. 3.
11. See James A. Greig, *Francis Jeffrey of the Edinburgh Review* (Edinburgh: Oliver and Boyd, 1948); Jonathan Cutmore, ed., *Conservatism and the Quarterly Review: A Critical Analysis* (London: Pickering & Chatto, 2007); George Lyman Nesbitt, *Benthamite Reviewing. The First Twelve Years of the Westminster Review, 1824–1836* (New York: Columbia University Press, 1933).
12. Paul A. Pickering and Alex Tyrell, *The People's Bread: A History of the Anti-Corn Law League* (London and New York: Leicester University Press, 2002); Adam Hochschild, *Bury the Chains: Prophets and Rebels in the Fight to Free and Empire's Slaves* (Boston: Houghton-Mifflin, 2005).
13. For examples from Germany, see Woodruff D. Smith, *Politics and the Sciences of Culture in Germany, 1840–1920* (New York: Oxford University Press, 1991), pp. 100–14, 162–73.
14. Ilja van den Broek, Christianne Smit, and Dirk Jan Wolffram, eds., *Commitment and Imagination: Representations of the Social Question* (Leuven: Peeters, 2010); Andrew Robert Aisenberg, *Contagion: Disease, Government, and the Social Question in Nineteenth-Century France* (Stanford: Stanford University Press, 1999).
15. Habermas, *Structural Transformation*, esp. pp. 37, 175–76, and Woodruff D. Smith, *Consumption and the Making of Respectability, 1600–1800* (New York and London: Routledge, 2002), pp. 139–70, 218–21, 234–35.
16. This phenomenon is described in a nineteenth-century American context in David T. Z. Mindich, *Just the Facts: How "Objectivity" Came to Define American Journalism* (New York: New York University Press, 2000), pp. 15–39.
17. See Lawrence E. Klein, *Shaftesbury and the Culture of Politeness: Moral Discourse and Cultural Politics in Early Eighteenth-Century England* (Cambridge: Cambridge University Press, 1994).
18. Then, as now, some writers could cultivate a reputation for intemperance and unreasonableness and be published on that account, although it might take them a while before they found the right niche. One example is Thomas Carlyle. See his *Chartism*, 2nd ed.(London: J. Fisher, 1840).
19. On Macaulay, see Clive, *Macaulay*. On Guizot, see Larry Siedentop's introduction to François Guizot, *The History of Civilization in Europe* (London: Penguin, 1997), pp. vii–xxxvii.
20. Habermas, *Structural Transformation*, pp. 88–117.

21. See Amanda Claybaugh's introduction to Harriet Beecher Stowe, *Uncle Tom's Cabin* (New York: Barnes & Noble Classics, 2003), pp. xiii–xlii.
22. See Ian Heywood, *The Revolution in Popular Literature: Print, Politics, and the People, 1790–1860* (Cambridge: Cambridge University Press, 2004), and Virginia Berridge, "Popular Sunday Papers and Mid-Victorian Society," in George Boyce, J. Curran, and P. Wingate, eds., *Newspaper History from the Seventeenth Century to the Present Day* (London: Constable, 1978), pp. 247–64. See also Simon Eliot and Jonathan Rose, eds., *A Companion to the History of the Book* (Malden, MA: Blackwell, 2007), especially the following articles: Rob Banham, "The Industrialization of the Book," pp. 273–90; Simon Eliot, "From Few and Expensive to Many and Cheap: The British Book Market 1800–1890," pp. 291–302; and Jean-Yves Mollier and Marie-Françoise Cachin, "A Continent of Texts: Europe 1800–1890," pp. 303–14.
23. For an explication of this model, see Henry Adams's description of the place of the *North American Review* in American journalism in the 1860s in Henry Adams, *The Education of Henry Adams* (Oxford: Oxford University Press, 1999), p. 198.
24. Albrecht Koschnik, "The Democratic Societies of Philadelphia and the Limits of the American Public Sphere, circa 1793–1795," *William and Mary Quarterly* 58, 3 (July 2001): 615–36; Michael Warner, *The Letters of the Republic: Publication and the Public Sphere in Eighteenth-Century America* (Cambridge, MA: Harvard University Press, 1990).
25. Thomas Bender, *New York Intellect: A History of Intellectual Life in New York City, from 1750 to the Beginnings of Our Own Time* (New York: Knopf, 1987), pp. 46–88.
26. Marshall Foletta, *Coming to Terms with Democracy: Federalist Intellectuals and the Shaping of an American Culture* (Charlottesville: University of Virginia Press, 2001).
27. Ibid.; Curtiss S. Johnson, *Politics and a Belly-full: The Journalistic Career of William Cullen Bryant, Civil War Editor of the New York Evening Post* (New York: Vantage Press, 1962).
28. See Robert E. Spiller, ed., *The Roots of National Culture* (New York: Macmillan, 1933).
29. See Sheldon D. Pollack, *War, Revenue, and State Building: Financing the Development of the American State* (Ithaca: Cornell University Press, 2009).
30. Testimony to the centering of public discourse in the United States around discussions in British periodicals can be found through a brief survey of the contents of practically any quality American journal from the period. See, for example, "Mr. Greely at the British Anti-Slavery Meetings," *Living Age* 30, 379 (August 28, 1851): 382–83, or John Stuart Mill, "The Contest in America," *Harper's New Monthly Magazine* 24, 143 (April 1862): 677–84.
31. Frances Trollope, *Domestic Manners of the Americans,* ed. Donald Smalley (New York: Alfred A. Knopf, 1949), pp. 43–48.
32. Ibid, pp. 156–57, 279–80, 404–9.
33. Alexis de Tocqueville, *Democracy in America,* ed. and abr. R. Heffner (New York: New American Library, 2001), p. 92.

34. George Wilson Pierson, *Tocqueville in America* (Baltimore and London: Johns Hopkins University Press, 1938), pp. 7–9.
35. For an indication of some of these perceptions, see the introductory statements in the first issues of *Harper's Monthly* and its companion, the weekly *International Magazine of Literature, Art, and Science: Harper's* 1, 1 (June 1850): 1–2; *International Magazine of Literature, Art, and Science* 1, 1 (July 1, 1850): 1–2.
36. On the founding of *Harper's*, see Eugene Exman, *The Brothers Harper: A Unique Publishing Partnership and Its Impact upon the Cultural Life of America from 1817 to 1853* (New York: Harpers, 1965), p. 304. See also Frank Luther Mott, *A History of American Magazines* (Cambridge: Cambridge University Press, 1948); Robert A. Gross, "Building a National Literature: The United States 1800–1890," in Eliot and Rose, eds., *Companion to the History of the Book*, pp. 315–28, esp. pp. 325–26.
37. See Sheila Post-Lauria, *Correspondent Colorings: Melville in the Marketplace* (Amherst: University of Massachusetts Press, 1996), pp. 151–209.
38. Henry Adams, surveying the daily newspapers of New York that exerted national influence in 1868 (the newspapers he had considered trying to work for), listed the *Times* (although not by name), the *Tribune*, the *Sun*, the *Herald*, and the *Evening Post*. H. Adams, *Education*, pp. 206–7.
39. Ibid., pp. 217–18.
40. For a description of several important members of this generation in the context of the times, see Louis Menand, *The Metaphysical Club* (New York: Farrar, Straus and Giroux, 2001).
41. Ibid., pp. 61–69.
42. H. Adams, *Education*, pp. 201–2.
43. See, for example, ibid., p. 233.
44. An example of this kind of thinking, from Henry Adams again: Henry Brooks Adams, "Civil Service Reform," *North American Review* 109, 225 (October 1869): 443–76.
45. See Henry Brooks Adams, "The Session," *North American Review* 117, 240 (July 1873): 204–6, and E. L. Godkin, "Commercial Immorality and Political Corruption," *North American Review* 107, 220 (July 1868): 248–67.
46. For example, Charles W. Eliot, "What Is a Liberal Education?" *Century Magazine* 28, 2 (June 1884): 203–12.
47. H. Adams, *Education*, pp. 203–4. Henry and Charles had exchanged letters on similar topics for quite some time. See letter from Henry Adams to Charles Francis Adams, Jr., February 9, 1859, in *Letters of Henry Adams (1858–1891)*, ed. by Worthington Chauncey Ford (Boston and New York: Houghton Mifflin, 1930), p. 20.
48. Charles Francis Adams, "A Chapter of Erie," *North American Review* 109 (July 1869): 30–106, and Henry Adams, "The New York Gold Conspiracy," *Westminster Review* (October 1870).
49. For an extremely interesting, admirably brief discussion of Henry Adams as journalist and professor, see Garry Wills, *Henry Adams and the Making of America* (Boston: Houghton-Mifflin, 2005), pp. 72–95.

50. H. Adams, *Education*, pp. 201–14, 217–19; Henry Brooks Adams, "The Session," *North American Review* 111, 228 (July 1870): 29–62, especially p. 62.
51. See Witold Rybczynski, *A Clearing in the Distance: Frederick Law Olmsted and America in the Nineteenth Century* (New York: Scribner, 1999); Laura Wood Roper, *Frederick Law Olmsted* (Baltimore: Johns Hopkins University Press, 1973); Linda Dowling, *Charles Eliot Norton: The Art of Reform in Nineteenth-Century America* (Lebanon, NH: University of New Hampshire Press, 2008); John T. Seaman, *A Citizen of the World: The Life of James Bryce* (London and New York: Tauris, 2006); Menand, *Metaphysical Club*, 73–148, 337–75.
52. On Godkin and the *Nation*, see William M. Armstrong, *E. L. Godkin and American Foreign Policy 1865–1900* (New York: Bookman Associates, 1957), pp. 11–37, and the editor's notations and chapter introductions in *The Gilded Age Letters of E. L. Godkin*, ed. William M. Armstrong (Albany: State University of New York Press, 1974). See also Bender, *New York Intellect*, pp. 181–91. Godkin's concern for the future of freed American slaves was not strong and his support for Reconstruction declined to essentially nothing by the mid-1870s.
53. On Bagehot, see the editor's introduction to Norman St. John Stevas, ed., *Bagehot's Historical Essays* (Garden City, NY: Doubleday & Co., 1965), pp. vii–xlii.
54. Gilman consulted Godkin in 1882 about the qualifications of the economist Richard T. Ely for a teaching position at Hopkins, believing that Ely was a regular contributor to the *Nation*. Letter Godkin to Gilman, June 1, 1882, *Gilded Age Letters*, p. 284. In 1884, Gilman allowed Abraham Flexner, then a 17-year-old undergraduate trying to get his degree in two years, to skip a required English composition course on the basis of correspondence Flexner had had published in the *Nation:* from Clark Kerr's introduction to Abraham Flexner, *Universities. American English German* (Oxford: Oxford University Press, 1965), p. viii.
55. James Bryce, *The American Commonwealth* (3 vols.; London and New York: Macmillan, 1888).
56. Ibid., 3: 3–63.
57. Ibid., 3: 426–64.
58. For example, see an issue of the *North American Review* published in the aftermath of the strikes of 1877. See especially "Fair Wages," solicited from "a Striker" not otherwise named, *North American Review* 125, 258 (September 1877): 322–27, and what amounts to a rebuttal by Thomas A. Scott, president of the Pennsylvania Railroad: "The Recent Strikes," *North American Review* 125, 258 (September 1877), pp. 351–63.
59. Armstrong, ed., *Gilded Age Letters*, pp. 195–96, 272.
60. Alexandra Villard De Borchegrave and John Cullen, *Villard: The Life and Times of an American Titan* (New York: Nan Talese, 2001).
61. Thomas L. Haskell, *The Emergence of Professional Social Science: The American Social Science Association and the Nineteenth-Century Crisis of Authority* (Urbana: University of Illinois Press, 1977), pp. 115–20.
62. Woodruff D. Smith, *The Ideological Origins of Nazi Imperialism* (New York: Oxford University Press, 1986), pp. 61–2.
63. See, for example, the views of Henry Adams about the need for a quality newspaper in Washington in "The Session," *North American Review* 117, 240

(July 1873): 204–6. In an earlier version of "The Session" (a title and concept that Adams lifted—with acknowledgment—from the British *Quarterly Review*), he had made it clear who would take part in the central conversations of the public sphere: "The American statesman or philosopher who would enter upon this great debate must make his appeal, not to the public opinion of a day or a nation, but to the minds of the few persons who, in every age and in all countries, attach their chief interest to the working out of the great problems of human society under all their varied conditions." Henry Brooks Adams, "The Session," *North American Review*, 111, 228 (July 1870): 62.
64. See "The Newspaper and the Organ," *Century* 29, 3 (January 1885): 461–62.
65. Pulitzer was, however, reported to be a regular reader of the *Post*. Armstrong, *Godkin and Foreign Policy*, p. 12.
66. See John Tebbel and Mary Ellen Zuckerman, *The Magazine in America 1741–1990* (New York: Oxford University Press, 1991), pp. 57–146.
67. Daniel T. Rodgers, *Atlantic Crossings: Social Politics in a Progressive Age* (Cambridge, MA: Belknap/Harvard University Press, 1998), pp. 33–75.
68. Bender, *New York Intellect*, pp. 121–30.
69. The principal study of the ASSA is Haskell, *Emergence of Professional Social Science*.
70. Dieter Lindenlaub, *Richtungskämpfe im Verein für Sozialpolitik: Wissenschaft und Sozialpolitik im Kaiserreich vornehmlich vom Beginn des Neuen Kurses bis zum Ausbruch des Ersten Weltkrieges, 1890–1914*, 2 vols. (Wiesbaden: Steiner, 1967).
71. See Peter Novick, *That Noble Dream. The "Objectivity Question" and the American Historical Profession* (Cambridge: Cambridge University Press, 1988).
72. Robert L. Beisner, *Twelve against Empire. The Anti-Imperialists, 1898–1900* (New York: McGraw-Hill, 1968). See also Armstrong, *Godkin*, pp. 162–99.
73. A. Lawrence Lowell, *Colonial Civil Service: The Selection and Training of Colonial Officials in England, Holland, and France, with an Account of the East India College at Haileybury (1806–1857) by H. Morse Stephens* (New York: Macmillan, 1900).
74. See Ari Hoogenboom, *Outlawing the Spoils: A History of the Civil Service Reform Movement* (Urbana: University of Illinois Press, 1961), and Beisner, *Twelve against Empire*.
75. This was recognized very clearly at the time. See "The Press and the New Reform," *Century* 25, 6 (April 1883): 954.
76. See, for example, George William Brown, "English Civil Service Reform," *Atlantic Monthly* 43, 259 (May 1879): 580–86.
77. This is beautifully described in the case of Oliver Wendell Holmes Jr. in Menand, *Metaphysical Club*, especially on pp. 43–44 and 59–60.
78. Ibid.
79. This is one of the main points made in "The Higher Education in America," *Galaxy* 11, 3 (March 1871): 369–86, a remarkably comprehensive overview of some of the main lines of university reform in the United States.
80. Ibid., p. 369; Eliot, "What is a Liberal Education?"
81. See Ted Curtis Smith, *The Gilded Age Press, 1865–1900* (Westport, CT: Praeger, 2003).

82. Michael Schudson, *Discovering the News: A Social History of American Newspapers* (New York: Basic Books, 1981), pp. 61–87.
83. Laurence R. Veysey, *The Emergence of the Modern American University* (Chicago: University of Chicago Press, 1970), pp. 57–179.
84. Although Bender, *New York Intellect*, pp. 206–62, argues that this happened at the expense of metropolitan cultures and discourses.
85. An example of an attempt to present a range of viewpoints can be seen in the 1877 issue of the *North American Review* on strikes cited previously: *North American Review* 125, 258 (September 1877): 322–27 and 351–63.
86. W. E. B. Du Bois published three articles in the *Atlantic Monthly*, although his biographer indicates that this was exceptional. David Levering Lewis, *W.E.B. Du Bois: Biography of a Race 1868–1919* (New York: Henry Holt, 1993), pp. 194, 197–201, 232, 262. For an example, see W. E. Borghardt Du Bois, "Strivings of the Negro People," *Atlantic Monthly* 80, 478 (August 1897): 194–98. The *Atlantic* was willing to regard African American intellectuals as legitimate participants in public discourse—at least as it concerned African American issues—although typically as presenters of a particular perspective among others in a series. See William A. Davis, "Reconstruction and Disfranchisement," *Atlantic Monthly* 88, 528 (October 1901): 433–99.

Chapter 4

1. The essay is conveniently available in Isaac Kramnick, ed., *The Portable Enlightenment Reader* (New York: Vintage, 1995), pp. 1–7.
2. Matthew Bernard Levinger, *Enlightened Nationalism. The Transformation of Prussian Political Culture, 1806–1848* (Oxford and New York: Oxford University Press, 2000), pp. 50–9.
3. Charles E. McClelland, *State, Society, and University in Germany 1700–1914* (Cambridge: Cambridge University Press, 1980), pp. 34–98.
4. Ibid., pp. 101–45.
5. Heinz Steinberg, *Wilhelm von Humboldt* (Berlin: Stapp, 2001).
6. Ernst Spranger, *Wilhelm von Humboldt und die Reform des Bildungswesens* (Tübingen: Max Niemeyer, 1960; orig. ed. 1910), pp. 133–240.
7. Ibid., pp. 199–210.
8. Humboldt's most general exposition of his ideas about university education, including the role of research, is his memorandum "Über die innere und äussere Organisation der höheren wissenschaftlichen Anstalten in Berlin," in Wilhelm von Humboldt, *Ausgewählte philosophische Schriften*, ed. Johannes Schubert (Leipzig: Felix Meiner, 1910), pp. 204–15.
9. A selection of excerpts from Humboldt's writings on education can be found in Wilhelm von Humboldt, *Humanist without Portfolio: An Anthology of the Writings of Wilhelm von Humboldt,* tr. Marianne Cowan (Detroit: Wayne State University Press, 1963), pp. 125–44.
10. See E.L. Godkin, "The Higher Education in America," *Galaxy* 11, 3 (March 1871): 371–2.

11. See Andrew Zimmerman, *Anthropology and Antihumanism in Imperial Germany* (Chicago: University of Chicago Press, 2001), pp. 1–14, 111–34.
12. See McClelland, *State, Society, University*, pp. 162–89, 211–17, 236–7, and Lenore O'Boyle, "Learning for Its Own Sake: The German University as a Nineteenth-Century Model," *Comparative Studies in Society and History*, 25, 1 (1983): 3–25.
13. This is demonstrated by a survey of the occurrence of the phrase "university reform" between 1840 and 1860 in the journals included in the online "Making of America" collection maintained by Cornell University Library (searched on November 8, 2007.) Practically all of the appearances between 1840 and 1860 refer to the issue in Britain (often in reprinted British articles). After 1860, most of the references are to the United States.
14. See Boyd Hilton, *A Mad, Bad, and Dangerous People?: England 1783–1846* (New York: Oxford University Press, 2006), pp. 599–611.
15. Emmelina Cohen, *The Growth of the British Civil Service 1780–1939* (London: Routledge, 1965), pp. 78–112; Richard Chapman, *The Civil Service Commission 1855–1971: A Bureau Biography* (London: Routledge, 2004), pp. 11–40.
16. See K. Theodore Hoppen, *The Mid-Victorian Generation 1846–1886* (New York: Oxford University Press, 2000), pp. 111–13, 497–98, 514.
17. See, for example, W. J. Ashley, "Jowett and the University Ideal," *Atlantic Monthly* 80, 477 (July 1897): 95–105.
18. See Evelyn Abbott and Lewis Campbell, *The Life and Letters of Benjamin Jowett*, 2 volumes (London: J. Murray, 1897), 2: 125, 129.
19. On the political importance of the constructed character of the "gentleman" in nineteenth-century British politics and society, see P. J. Cain and A. G. Hopkins, *British Imperialism: Innovation and Expansion, 1688–1914* (London and New York: Longman, 1993).
20. In an unsigned article in the *Nation* in 1876, a British correspondent describes the extent to which university graduates have occupied significant positions throughout the structures of public life, principally as a result of the reforms in university education since 1850. *Nation* 22, 563 (April 13, 1876): 242–43. The article appears to be included because the editor thought it relevant to American concerns of the period.
21. This was the model, for example, that lay behind A. Lawrence Lowell's previously cited suggestions for organizing an American colonial service in 1899: Lawrence Lowell, *Colonial Civil Service: The Selection and Training of Colonial Officials in England, Holland and France, with an Account of the East College at Haileybury (1806–1857) by H. Morse Stephens* (New York: Macmillan, 1900). As president of Harvard, Lowell was primarily responsible for the attempt in the late 1920s and 1930s to reconstruct undergraduate education at Harvard along something like Oxford and Cambridge lines.
22. Thomas Balogh, "The Apotheosis of the Dilettante: The Establishment of the Mandarins," in Hugh Thomas, ed., *Crisis in the Civil Service* (London: Anthony Blond, 1968), pp. 11–52.

23. Sir Sydney Caine, *The History of the Foundation of the London School of Economics and Political Science* (London: London School of Economics and Political Science, 1963).
24. The standard modern book on the subject is Veysey, *Emergence of the Modern American University*.
25. Mark R. Nemec, *Ivory Towers and Nationalist Minds: Universities, Leadership, and the Development of the American State* (Ann Arbor: University of Michigan Press, 2006).
26. Michael Dennis, *Lessons in Progress: State Universities and Progressivism in the New South, 1880–1920* (Urbana: University of Illinois Press, 2000).
27. Julie Reuben, *The Making of the Modern University* (Chicago: University of Chicago Press, 1996).
28. Thomas Bender, *Intellect and Public Life. Essays on the Social History of Academic Intellectuals in the United States* (Baltimore: Johns Hopkins University Press, 1993).
29. Allan Nevins, *The State Universities and Democracy* (Urbana: University of Illinois Press, 1962), p. 26.
30. Jefferson's mature thinking on universities is laid out in the report he wrote in 1818 as chair of a committee to propose a state university for Virginia: "Report of the Commissioners for the University of Virginia," in Thomas Jefferson, *Writings*, ed. Merrill D. Peterson (New York: Literary Classics of the United States, 1984), pp. 457–73.
31. See, among a great many examples, Charles W. Eliot, "What is a Liberal Education?" *Century Magazine* 28, 2 (June 1884): 203–12; Daniel Coit Gilman, "Education in America," *North American Review* 122 (January 1876): 191–228; C. K. Adams, "Ought the State to Provide for Higher Education?" *New Englander* 37, 141 (May 1878): 364–65; Noah Porter, *The American Colleges and the American Public* (New Haven: Charles Chatfield, 1870), which is an expansion of a series of articles by Porter in the *New Englander* in 1869.
32. See John Spencer Clark, *The Life and Letters of John Fiske*, Vol. 1 (Honolulu: University Press of the Pacific, 2006).
33. [John Fiske] "Considerations on University Reform," *Atlantic Monthly* 19, 114 (April 1867): 451–65. The author's name was not given with the article, but it came to be known quickly (a kind of game played by writers and readers in the quality journals before bylines became standard) and is listed in the *Atlantic's* index.
34. See, for example, a letter to the *Nation* signed "W.P.A." responding (favorably) to the article: *Nation* 4, 96 (May 2, 1867): 359–60.
35. James is placed in his public context in Menand, *Metaphysical Club*, pp. 73–95, 117–48, and passim.
36. H. Adams, *Education*, pp. 244–63.
37. See, for example, the assessment of Wills, *Henry Adams*, pp. 87–96. Adams himself even admitted in the *Education* that others, including President Eliot, thought he had done a good job. H. Adams, *Education*, p. 256.
38. H. Adams, *Education*, p. 246.

39. "The Higher Education in America," pp. 370–71.
40. The relationship between public discourse and history in the minds of reformers at Harvard is suggested by the fact that Adams's immediate predecessor as editor of the *North American Review* was one of the first two people appointed to teach history at Harvard: Ephraim Whitman Gurney. According to Adams, it was Gurney who decided that Adams was the ideal person to take over the editorship. H. Adams, *Education*, p. 246. In 1870, Eliot had offered a chair in history to E. L. Godkin of the *Nation* before he offered a position to Adams. Armstrong, ed., *Gilded Age Letters*, pp. 149–50.
41. Letter from Henry Adams to Charles Milnes Gaskell, September 29, 1870, in *Letters of Henry Adams*, p. 194.
42. In the *Education*, Adams connects this emphasis to both his acceptance of German academia as the world's most advanced example of higher education and the political prestige acquired by the newly united German Empire in its defeat of France in 1870–71. "Germany was never so powerful, and the Assistant Professor of History had nothing else as his stock in trade. He imposed Germany on his scholars with a heavy hand. He was rejoiced; but he sometimes doubted whether they should be grateful." H. Adams, *Education*, p. 255.
43. Seminars directly modeled on those of Berlin University were introduced at the University of Michigan by its president, Philip Henry Tappan, in the 1850s, but they do not appear to have caught on there. See James Turner and Paul Bernard, "The German Model and the Graduate School: The University of Michigan and the Foundation Myth of the American University," in Roger L. Geiger, ed., *The American College in the Nineteenth Century* (Nashville: Vanderbilt University Press, 2000), pp. 221–41.
44. Wills, *Henry Adams*, pp. 87–92.
45. See Ernest Samuels, *Henry Adams: The Middle Years* (Cambridge, MA: Belknap Press of Harvard University Press, 1958).
46. Henry Brooks Adams, "The Session," *North American Review* 111, 228 (July 1870): 29–62, especially p. 62. Adams makes a connection between the need for a high level of intellect in the government service and the need for a high-class newspaper in Washington in "The Session," *North American Review* 117, 240 (July 1873): 204–6. See also E. L. Godkin, "Commercial Immorality and Political Corruption," *North American Review* 107, 220 (July 1868): 248–67, and A. R. Macdonough, "Civil Service Reform," *Harper's New Monthly Magazine* 40, 238 (March 1870): 546–56, especially p. 548.
47. Nemec, *Ivory Towers*, pp. 77–108.
48. Letter Godkin to Gilman, June 1, 1882, *The Golden Age Letters of E.L. Godkin*, ed. William M. Armstrong (Albany: State University of New York Press, 1974), p. 284.
49. Jefferson, *Writings*, pp. 275, 458, 472, 475, 1207–8.
50. See Richard E. Abel and Lyman W. Newlin, eds., *Scholarly Publishing: Books, Journals, Publishers and Libraries in the Twentieth Century* (New York: Wiley, 2002).

51. Peter Givler, "University Press Publishing in the United States," in Abel and Newlin, eds., *Scholarly Publishing,* pp. 108–20, and on the website of the American Association of University Presses at www.aaupnet.org.
52. George Santayana, *Character and Opinion in the United States, with Reminiscences of William James and Josiah Royce and Academic Life in America* (New York: Charles Scribner's Sons, 1920), pp. 61–3, 91–3.
53. Josiah Royce, "Present Ideals of American University Life," *Scribner's Magazine* 10, 3 (September 1891): 384. Note that Royce is making this point in a quality general-interest magazine.
54. It should be noted that several of the leading founders of the American public sphere were in reality quite intolerant of views that differed from their own— E. L. Godkin perhaps most notoriously. See Armstrong, ed., *Gilded Age Letters,* pp. 473–4.
55. This is the argument, for example, of Charles W. Eliot in his article "What Is a Liberal Education?" in *Century Magazine* 28, 2 (June 1884): 203–12.
56. See Veysey, *Emergence*; Nemec, *Ivory Towers*; Reuben, *Making of the Modern University.*
57. The primary text for the survey will be the already cited anonymous article "The Higher Education in America," *Galaxy* 11, 3 (March 1871): 369–86, which is unusually comprehensive and is itself a critique of a book by Noah Porter, the principal defender of the traditional American college and opponent of university reform. (Porter, *The American Colleges.*) Other points of view will be discussed and cited mainly in reference to the *Galaxy* article. The author appears to be a member of the Columbia faculty—one who clearly thinks that Columbia is far behind the leaders in adopting changes that he favors.
58. David B. Potts, "Curriculum and Enrollment: Assessing the Popularity of Antebellum Colleges," in Geiger, ed., *American College,* pp. 37–45.
59. "Higher Education in America," pp. 372–73.
60. A major defense of the traditional college (although it accepts the notion that some things could be usefully modified) is Porter, *American Colleges.*
61. "German Universities," *Atlantic Monthly* 7, 41 (March 1861): 257–72.
62. Ibid., pp. 266–7.
63. Porter, *American Colleges,* pp. v–vi.
64. "Higher Education in America," p. 369.
65. Ibid.
66. Quoted in ibid., p. 370.
67. Ibid., p. 371.
68. Ibid.
69. Eliot, "What Is a Liberal Education?"
70. On Tappan, see Charles M. Perry, *Philip Henry Tappan* (Ann Arbor: University of Michigan Press, 1933), and Bender, *New York Intellect,* pp. 108–14. Eliot and his supporters carried through reforms at Harvard much more thoroughly than Tappan, who was fired in 1863, was able to do at Michigan.
71. See, for example, an article entitled "The Course at Harvard," *New York Times,* April 8, 1883, which states that Harvard was "the pioneer on this continent" of

the elective system "and it remains today its leading exponent." The first part of the statement was true in 1883 only if "elective system" refers to the extreme version of the concept, but since the article divides current approaches to curriculum into two categories—the old one, based on prescribed studies, and the new one, based on electives—the term must refer to curricula featuring electives to any substantial extent. And as the article points out, even Harvard retained a prescribed first-year curriculum in the 1880s.

72. Vesey, *Emergence of the Modern American University,* p. 68.
73. [Fiske], "Considerations," pp. 459–61.
74. See the anonymous article "The American College," *Atlantic Monthly,* 75, 451 (May 1895): 703–7. The article is quite critical of the practical results of university reform, although not from the position of a die-hard defender of the traditional college. It accepts the college as the "natural" format for higher education in America, which needs to be improved rather than replaced by other kinds of institution.
75. "Higher Education in America," p. 375.
76. Ibid., pp. 383–84.
77. James C. Albisetti, "German Influence on the Higher Education of American Women, 1865–1914," in Henry Geitz, Jürgen Heideking, and Jurgen Herbst, eds., *German Influences on Education in the United States to 1917* (Cambridge: German Historical Institute, Washington, D.C., and Cambridge University Press, 1995), 238–40.
78. Ibid., pp. 227–44, esp. 235–36.
79. See, for example, Kate Gannett Wells, "Women in Organizations," *Atlantic Monthly* 46, 275 (September 1880): 360–68, in which the role of women in the public sphere is discussed by a prominent anti-suffragist contributor to the quality magazines.
80. "Higher Education in America," p. 373.
81. Ibid., pp. 373, 378–80.
82. Daniel Coit Gilman, *University Problems in the United States* (New York: Garrett Press, 1969; reprint of 1898 edition) pp. 171–72.
83. D.C. Gilman, "The Future of American Colleges and Universities," *Atlantic Monthly* 78, 466 (August 1896): 178.
84. This point is made in detail in C. K. Adams, "Ought the State to Provide for Higher Education?" *New Englander* 37, 141 (May 1878): 370–84.
85. See ibid., pp. 362–84, and George E. Howard, "The State University in America," *Atlantic Monthly* 67, 401 (March 1891): 332–42.
86. See [F. H. Hedge], "University Reform. An Address to the Alumni of Harvard, at their Triennial Festival, July 19, 1866," *Atlantic Monthly* 18, 107 (September 1866): 296–307, especially p. 299, and "Higher Education in America," p. 371.
87. The strategy is most comprehensively presented in Howard, "State University."
88. At least the rapid rise in the number of tiny private colleges came to an end. See the summary of the decennial federal count of institutions in the United States classified as "colleges" or "universities" (173 in 1840; 234 in 1850; 467 in 1860; 426 in 1871; 591 in 1880; 451 in 1890; and 530 in 1902) in Seymour

E. Harris, *A Statistical Portrait of Higher Education* (New York: McGraw-Hill, 1972), p. 924. The later numbers include a substantial number of new public institutions established in the last quarter of the nineteenth century.
89. See Veysey, *Emergence of the American University,* pp. 98–113.
90. For an exposition of this view, see James Baldwin Turner, "Plan for an Industrial University, for the State of Illinois," a document prepared for an 1851 convention, in Theodore Rawson Crane, ed., *The Colleges and the Public* (New York: Columbia University Teachers College, 1961), pp. 172–89.
91. Bender, *New York Intellect,* pp. 114–16.
92. Jefferson, *Writings,* pp. 457–73.
93. Populism has begun to be rehabilitated as a significant and positive contribution to the development of the modern United States. See, for example, Charles Postel, *The Populist Vision* (New York: Oxford University Press, 2007). A recent dissertation by Scott Gelber specifically addresses the nature, coherence, and effect of Populist programs for higher education and is the source for most of the discussion of this topic here: Scott M. Gelber, *Academic Populism: The People's Revolt and Higher Education, 1880–1905* (Harvard Ph.D. Dissertation, 2008).
94. See, for example, Turner, "Plan for an Industrial University," pp. 172–89.
95. For the role of the Carnegie Foundation in some of these developments and for the motives behind them, see Ellen Condliffe Lagemann, *Private Power for the Public Good: A History of the Carnegie Foundation for the Advancement of Teaching* (Middletown, CT: Wesleyan University Press, 1983), pp. 94–121.
96. These developments are summarized in Christopher J. Lucas, *American Higher Education: A History* (New York: St. Martin's Griffin, 1994), esp. pp. 200–4. See also Frederick Rudolph, *The American College and University: A History* (New York: Alfred A. Knopf, 1968), pp. 462–82.
97. This point was made in the 1950s by James Bryant Conant, former president of Harvard. See James Bryant Conant, *The Citadel of Learning* (New Haven: Yale University Press, 1956), p. 24.

Chapter 5

1. Seymour E. Harris, *A Statistical Portrait of Higher Education* (New York: McGraw-Hill, 1972), p. 945.
2. Ibid., pp. 926–27.
3. Christopher J. Lucas, *American Higher Education: A History* (New York: St. Martin's Griffin, 1994), p. 230.
4. This has changed in recent years, due in large part to the adoption of something like American views of the nature and function of higher education in other countries. By 2006, the United States had fallen from its long-time position of first to seventh place among countries belonging to the Organisation for Economic Co-Operation and Development in the proportion of its population attaining tertiary education. See *Education at a Glance 2006,* available on the OECD website at www.oecd.org/dataoecd/51/20/37392850.pdf.

5. See enrollment statistics derived from annual federal reports available online from the National Center for Educational Statistics of the United States Department of Education at http://nces.ed.gov.
6. In 1902, there were 530 colleges and universities in the United States, most of them private. (Harris, *Statistical Portrait*, p. 924.) Fewer than 40 percent of students attended public institutions. In 2004, there were about 4,200 colleges and universities—630 of them public four-year and graduate institutions and around 1,100 of them public two-year institutions. The public four-year and graduate institutions enrolled 6.2 million students; their private counterparts enrolled 3.2 million. The public two-year colleges enrolled about 6 million students; private two-year enrollments were small. Peter D. Eckel and Jacqueline E. Krier, *An Overview of Higher Education in the United States: Diversity, Access, and the Role of the Marketplace* (Washington, D.C.: American Council on Education, 2004), pp. 1–2.
7. This process was particularly notable in medical education. See W. F. Bynum, "Sir George Newman and the American Way," in Vivian Nutton and Roy Porter, eds., *History of Medical Education in Britain* (Amsterdam and Atlanta: Rodopi, 1995), pp. 37–68, and William H. Schneider, ed., *Rockefeller Philanthropy and Modern Biomedicine: International Initiatives from World War I to the Cold War* (Bloomington: Indiana University Press, 2002).
8. Thomas Neville Bonner, *Iconoclast: Abraham Flexner and a Life in Learning* (Baltimore: Johns Hopkins University Press, 2002), pp. 69–90; Barbara Barzansky and Norman Gevitz, eds., *Beyond Flexner: Medical Education in the Twentieth Century* (Westport, CT: Greenwood, 1992).
9. See Mary Roth Walsh, *Doctors Wanted: No Women Need Apply* (New Haven: Yale University Press, 1977).
10. Robert Stevens, *Law School: Legal Education in America from the 1850s to the 1980s* (Chapel Hill: University of North Carolina Press, 1983).
11. Some of the pre-Flexner impetus to the growing connection between professionalism and universities can be seen in *Annals of the American Academy of Political and Social Science* 28, 1 (July 1906), which is entitled *The Business Professions* and devoted entirely to that subject. The principal theme of most of the articles (some by business leaders and some by social scientists) is the value of a university education for particular professions and the need to develop a place in higher education for professional business studies.
12. See Peter Novick, *That Noble Dream. The "Objectivity Question" and the American Historical Profession* (Cambridge: Cambridge University Press, 1988).
13. See, for example, the series of pamphlets published by the AHA during the Second World War to educate service personnel on issues having to do with the war and with the postwar world. These are available on the AHA website: www.historians.org/Projects/GIroundtable.
14. Daniel Coit Gilman, *University Problems in the United States* (New York: Garrett Press, 1969; reprint of 1898 edition), pp. 153–62; Mark R. Nemec, *Ivory Towers and Nationalist Minds: Universities, Leadership, and the Development of the American State* (Ann Arbor: University of Michigan Press, 2006), pp. 77–108.

15. Schneider, ed., *Rockefeller Philanthropy*; E. Richard Brown, *Rockefeller Medicine Men: Medicine and Capitalism in America* (Berkeley: University of California Press, 1979); Dwight Macdonald, *The Ford Foundation: The Men and the Millions* (New Brunswick, NJ: Transaction Publishers, 1989; orig. ed. 1956).
16. See Richard E. Abel and Lyman W. Newlin, eds., *Scholarly Publishing: Books, Journals, Publishers and Libraries in the Twentieth Century* (New York: Wiley, 2002).
17. See the website of the American Association of University Presses: aaupnet.org.
18. A recent case was the dismissal of Ward Churchill from a tenured position at the University of Colorado—technically because of scholarly misconduct but clearly in response to public outcry against his statements about the events of September 11, 2001. See John Gravois, "Colo. Regents Vote to Fire Ward Churchill," *Chronicle of Higher Education* 53, 48 (August 3, 2007): 1.
19. An example: an op-ed piece in the *Boston Globe,* January 1, 2006, by James Lang entitled "A Vote for Tenure. Why the Tenure System, Despite the Popular Caricatures, Is Essential to Higher Education." Nowhere in the article is the function of protecting freedom of speech, "academic freedom," or "intellectual freedom" mentioned—either as a reason that the public should support tenure or in any other context.
20. This is a central part of the argument for tenure given by James O. Freedman. I do not find it particularly compelling, and I suspect that much of the public at large would also find it less than convincing. See James O. Freedman, *Liberal Education and the Public Interest* (Iowa City: University of Iowa Press, 2003), pp. 40–43.
21. See Thomas Bender, *New York Intellect: A History of Intellectual Life in New York City, from 1750 to the Beginnings of Our Own Time* (New York: Knopf, 1987), pp. 176–91.
22. See, for example, Edwin L. Godkin, *Unforeseen Tendencies of Democracy* (Freeport, NY: Books for Libraries Press, 1971; reprint of 1898 edition).
23. On American Populism, see Charles Postel, *The Populist Vision* (New York: Oxford University Press, 2007).
24. This point is important to Jürgen Habermas's account of what happened to the public sphere in the nineteenth century: Jürgen Habermas, *The Structural Transformation of the Public Sphere: An Inquiry into a Category of Bourgeois Society* (Cambridge, MA: MIT Press, 1989), pp. 159–222.
25. William M. Armstrong, *Godkin and Foreign Policy 1865-1900* (New York: Bookman Associates, 1957), p. 12.
26. Thomas Woody, *History of Women's Education in the United States* (Lancaster, PA: Science Press, 1929), pp. 140–50.
27. The turning point came in the late 1970s, at both public and private institutions. See National Center for Educational Statistics, *Digest of Educational Statistics 2000,* Tables 172 and 173, available at nces.ed.gov.
28. Daryl G. Smith, "Women's Colleges and Coed Colleges: Is There a Difference for Women?" *Journal of Higher Education* 61 (1990): 181–95.

29. See Anthony Giddens, *The Third Way. The Renewal of Social Democracy* (Cambridge: Polity Press, 1998). Ultimately, as Giddens points out, social reality is composed of individuals and individual actions, but individuals must be placed in groups to be understood as social actors, and they must place themselves in groups in order to act.
30. One example out of many that could be cited: Kate Gannett Wells, a prominent Boston intellectual and anti-suffragist, published five full articles in the *Atlantic Monthly* between 1880 and 1885. *Atlantic Monthly* 46, 275 (September 1880); 46, 278 (December 1880); 48, 290 (December 1881); 54, 322 (August 1884); 55, 332 (June 1885).
31. Jean H. Baker, *Sisters: The Lives of America's Suffragists* (New York: Hill and Wang, 2005); Kathleen Barry, *Susan B. Anthony: A Biography of a Singular Feminist* (New York: New York University Press, 1988); Harriet Sigerman, *Elizabeth Cady Stanton: The Right is Ours* (New York: Oxford University Press, 2001). Not all suffragists did this, however. For one who sought a wider audience through more sensational publicity, see Amanda Frisken, *Victoria Woodhull's Sexual Revolution: Political Theater and the Popular Press in Nineteenth-Century America* (Philadelphia: University of Pennsylvania Press, 2004).
32. Elizabeth Silverthorne, *Sarah Orne Jewett: A Writer's Life* (Woodstock, NY: Overlook Press, 1993); Josephine Donovan, *Sarah Orne Jewett*, rev. ed. (Christchurch, NZ: Cybereditions, 2001).
33. One of her novels, for example, was *A Country Doctor* (1884). The "country doctor" in question was a woman, and the book sympathetically portrayed the difficulties faced by women professionals. Sarah Orne Jewett, *A Country Doctor* (New York: Bantam, 1999).
34. Shari Benstock, *No Gifts from Chance: A Biography of Edith Wharton* (New York: Scribner's, 1994), pp. 3–48.
35. Louise W. Knight, *Citizen: Jane Addams and the Struggle for Democracy* (Chicago: University of Chicago Press, 2005), pp. 80–108.
36. Paula J. Giddings, *Ida: A Sword among Lions. Ida B. Wells and the Campaign against Lynching* (New York: HarperCollins, 2008), pp. 15–39.
37. James Woodress, *Willa Cather: A Literary Life* (Lincoln: University of Nebraska Press, 1989), pp. 64–88.
38. Ibid., pp. 184–212.
39. See Jan Whitt, *Women in American Journalism: A New History* (Urbana: University of Illinois Press, 2008), for an overview of the subject with regard to journalists. The educational backgrounds of women journalists appear to have been quite varied, although as the twentieth century went on, college became standard.
40. My mother's one long text was a contribution on the Army Nurse Corps in Europe to the army's official history of the Second World War, which presumably sits in typescript in some archive of the Department of Defense to this day.
41. Or perhaps more accurately, what Christopher Newfield calls the "mass middle class" and that he describes as being primarily a product of the public universities. Christopher Newfield, *Unmaking the Public University: The Forty-year*

Assault on the Middle Class (Cambridge, MA: Harvard University Press, 2008), pp. 19–30.
42. David Levering Lewis, *W. E. B. DuBois: Biography of a Race 1868–1919* (New York: Henry Holt, 1993), pp. 261–73.
43. Christopher J. Lucas, *American Higher Education: A History* (New York: St. Martin's Griffin, 1994), pp. 207–8.
44. Raymond Arsenault, *Freedom Riders: 1961 and the Struggle for Racial Justice* (New York: Oxford University Press, 2006), pp. 533–87. Arsenault did not calculate the numbers of university students or classify their schools. I arrived at those figures by going through all of the background material assembled by Arsenault in the appendix to his book.
45. David Halberstam, *The Children* (New York: Fawcett, 1998), pp. 63–66, 73–76, 150–53, 156–58.
46. Vanessa Murphree, *The Selling of Civil Rights: The Student Nonviolent Coordinating Committee and the Use of Public Relations* (New York: Routledge, 2006).
47. Ibid., pp. 109–30.
48. Joel M. Roitman, *The Immigrants, the Progressives, and the Schools: Americanization and the Impact of the New Immigration upon Public Education in the United States* (Stark, KS: De Young Press, 1996).
49. Richard A. Hogarty, Aundrea E. Kelley, and Robert C. Wood, *Turnabout Time: Public Higher Education in the Commonwealth* (Boston: McCormack Institute, University of Massachusetts Boston, 1995).
50. Sydney C. Van Nort, *The City College of New York* (Charleston, SC: Arcadia Publishing, 2007).

Chapter 6

1. Lippmann is the subject of an outstandingly intelligent and comprehensive biography: Ronald Steele, *Walter Lippmann and the American Century* (Boston: Atlantic Monthly-Little, Brown, 1980).
2. Ibid., pp. 3–73.
3. Ibid., pp. 101–15.
4. The division among public intellectuals during the First World War is a major topic of American cultural history. See, for example, Steele, *Lippmann*, pp. 96–127; Thomas Bender, *New York Intellect: A History of Intellectual Life in New York City, from 1750 to the Beginnings of Our Own Time* (New York: Knopf, 1987), pp. 296–302; and Robert Westbrook, *John Dewey and American Democracy* (Ithaca: Cornell University Press, 1991), pp. 195–227.
5. Steel, *Lippmann*, pp. 116–54.
6. Ibid., pp. 155–70.
7. Ibid., pp. 197–219, describes Lippmann's relationship to the *World* and some of the paper's background. See also James R. Barrett, *Joseph Pulitzer and His World* (New York: Vanguard, 1941).

8. Walter Lippmann, *Public Opinion* (New York: Free Press, 1965; original edition 1922); Walter Lippmann, *The Phantom Public* (New York: Harcourt, Brace, 1925).
9. Lippmann, *Public Opinion*, pp. 191–252; Lippmann, *Phantom Public*, pp. 22–39.
10. Lippmann, *Phantom Public*, pp. 81–94.
11. Ibid., pp. 13–53; Lippmann, *Public Opinion*, pp. 58–64, 118–19, 170–90, 272–75.
12. See Lippmann, *Public Opinion*, pp. 251–75.
13. Ibid., pp. 191–233.
14. Ibid., pp. 56–64; Lippmann, *Phantom Public*, pp. 16–21.
15. Lippmann, *Public Opinion*, pp. 77–128, 157–69, 234–49.
16. Ibid., pp. 68–76, 157–69.
17. Ibid., pp. 367–97.
18. Lippmann, *Phantom Public*, pp. 40–53.
19. Lippmann, *Public Opinion*, pp. 411–18.
20. Lippmann, *Phantom Public*, pp. 22–39.
21. Ibid., pp. 54–62, 95–106, 144–45. Lippmann does suggest that the members of the self-interested groups into which the public is divided can be taught to recognize when somebody else's interests are being disguised behind a mask of supposed objectivity, when science and expertise are clearly being misused. They might also, if the methodology for doing so were to be developed by social scientists, learn to tell whether a policy that is proposed or has been adopted is rational—with "rationality" understood as an attempt to follow or establish fixed, regular rules and irrationality defined as arbitrariness. This is not very satisfactory. Apparently, if the state were to decide to kill its Jewish citizens, that would be acceptable if it were made a regular procedure and applied generally, rather than undertaken to serve particular interests and applied intermittently: Heinrich Himmler's view of the Holocaust rather than Hermann ("*I* decide who is Jewish") Göring's.
22. In *Public Opinion*, Lippmann does discuss newspapers, but he treats them as suppliers of news—that is, "facts" the reliability of which is rendered questionable by the stereotypes in the heads of reader-customers and those in the heads of the newspaper people themselves, as well as the interests of the people who own and manipulate the newspapers. He pays little attention to the discussion of ideas in the press and essentially none to the functions of periodicals such as the *New Republic*. Lippmann, *Public Opinion*, pp. 317–65.
23. John Dewey, *The Public and Its Problems* (Athens, OH: Swallow Press—Ohio University Press, 1991; orig. ed. 1927), pp. 116–17.
24. Westbrook, *John Dewey*, pp. 150–194.
25. Dewey, *Public and Its Problems*, pp. 12–28.
26. Ibid., pp. 15–16.
27. Ibid., pp. 69–74.
28. Ibid., p. 111.
29. Ibid., pp. 110–42.

30. Dewey clearly intends his book to be read by interested and educated members of the public as a contribution to a national conversation. It is not a face-to-face conversation, but readers can certainly talk about the book with each other, should they choose to do so. Moreover, the book itself is derived from a series of lectures Dewey gave at Kenyon College in 1926. The academic setting provided him a live audience that the published book extended more widely. He obviously knows this, but it has no effect on his analysis of the public and its discourse.
31. Dewey, *Public and its Problems,* pp. 123–25, 136–37, 203–9.
32. Ibid., pp. 96–98, 143–84.
33. John Dewey, *Democracy and Education: An Introduction to the Philosophy of Education* (New York: Free Press, 1944; orig. ed. 1916).
34. Westbrook, *John Dewey,* pp. 150–94.
35. Abraham Flexner, *Universities. American English German* (Oxford: Oxford University Press, 1965), p. xii.
36. Ibid., p. xv.
37. Ibid., p. 9.
38. Ibid., p. 10.
39. See the editorial (presumably by Godkin) entitled " 'Educated Men' in Centennial Politics," *Nation* 22, 575 (July 6, 1876): 5–6, which makes exactly this point.
40. Flexner, *Universities,* p. 30.
41. Daniel Coit Gilman, *University Problems in the United States* (New York: Garrett Press, 1969; reprint of 1898 edition), pp. 163, 171–72.
42. See David Levering Lewis, *W. E. B. DuBois: Biography of a Race 1868–1919* (New York: Henry Holt, 1993), pp. 549–51, which describes a 1915 meeting of the General Education Board, an arm of the Rockefeller Foundation, to decide on priorities for educational development for African Americans in the South. Lewis's account is based on minutes found in the Rockefeller Foundation archives. Flexner is cited as saying that it had been a mistake to give women access to the same forms of higher education that had been designed for men. It was important that the same mistake not be made with regard to African Americans—by which he clearly meant African Americans in general, not just African American women.
43. Flexner, *Universities,* p. 11.
44. Ibid., pp. 52–54.
45. Ibid., pp. 46–52.
46. If one accepts the belief that there is a single scale of fitness for university study (reducible, in most versions, to uniform subscales of preparation and intelligence), ones accepts also the basis of the argument that people who score higher on whatever instrument is used to measure individuals against the scale ought to be admitted to the university in preference to those who score lower. From this standpoint, "affirmative action" admissions, in which race or other background factors are taken into account, are deviations, suspect even though there may be pressing social reasons to make them. Academics who favor some measure

of affirmative action but who subscribe (usually uncritically) to the academic ideology are left in a quandary. Others have rejected affirmative action on the same grounds. The fact that the academic ideology legitimates the concept of a single qualification scale has given the political attack on affirmative action great force. If the ideology accepted the idea that universities actually have multiple functions, constituencies, and student bodies, fewer problems would present themselves. Affirmative action is largely consistent with the reality of a modern university with multiple roles in contemporary society. Opposition to it on grounds of "standards" and "fairness" is based on a fiction, but that fiction is firmly entrenched in the ideology that has echoed "down the decades" from Flexner.

47. The principal biographies of Hutchins are Harry S. Ashmore, *Unseasonable Truths: The Life of Robert Maynard Hutchins* (Boston: Little, Brown and Co., 1989) and Mary Ann Dzuback, *Robert M. Hutchins: Portrait of an Educator* (Chicago: University of Chicago Press, 1991).
48. Robert Maynard Hutchins, *Higher Learning in America* (New Haven and London: Yale University Press, 1936; reissued 1962 with a new introduction).
49. Ibid., p. 2.
50. Ibid., pp. 4–13, 33.
51. Ibid., p. 95.
52. Ibid., p. 107.
53. Ibid., pp.109–10.
54. Ibid., pp. 59–87.
55. Ibid., pp. xiv–xv.
56. "Democracy is the best form of government," Hutchins wrote—admittedly at the height of the Second World War. He went on immediately to say that democracy can be realized if we grasp the principles on which it rests, which he proceeded to postulate in quite idiosyncratic terms. Robert Maynard Hutchins, *Education for Freedom* (Baton Rouge: Louisiana State University Press, 1943), p. 95.
57. Hutchins, *Higher Learning*, pp. 13–21, 59–87.
58. Stephen Bates, *Realigning Journalism with Democracy: The Hutchins Commission, Its Times, and Ours* (Washington, D.C.: Annenberg Washington Program in Communication Studies, 1995).
59. John C. Nerone, *Last Rites: Revisiting Four Theories of the Press* (Urbana: University of Illinois Press, 1995), pp. 77–100. The report itself can be found in Commission on Freedom of the Press, *Free and Responsible Press*.
60. Commission on Freedom of the Press, *Free and Responsible Press*, pp. 20–29. (These are the basic "requirements" for a socially responsible press, elaborated briefly on the pages cited.)
61. The principal biography of Conant is James Hershberg, *James B. Conant: Harvard to Hiroshima and the Making of the Nuclear Age* (Stanford: Stanford University Press, 1995), which focuses on the subject of its subtitle but which deals thoroughly with other aspects of Conant's career. Conant wrote an autobiography:

James Bryant Conant, *My Several Lives: Memoirs of a Social Inventor* (New York: Harper & Row, 1970).
62. James Bryant Conant, *The Citadel of Learning* (New Haven: Yale University Press, 1956).
63. Ibid., pp. 1–22.
64. Conant's commitment to breadth of discussion and to toleration had decided limits, as was indicated by his tendency, while president of Harvard, to reject left-wing faculty for tenure. His prejudice was not, however, absolute, and within his limits, he seems to have made a conscientious effort to insist on intellectual toleration in academia. See the description of Conant's bumpy relationship with John Kenneth Galbraith in Glenn, "John Kenneth Galbraith," *Chronicle of Higher Education,* May 1, 2006. Nevertheless, Conant adopted the position that, at least for the duration of the Cold War, full academic freedom should not be granted to communists on the grounds that they (not individually, but taken as a category) do not themselves subscribe to academic freedom for others. Apparently, you only have to afford freedom of expression to those who can be presumed to agree with you that free speech is a good idea. See Wilson Smith and Thomas Bender, *American Higher Education Transformed, 1940–2005* (Baltimore: Johns Hopkins University Press, 2008), p. 458.
65. Conant, *Citadel of Learning,* p. 25.
66. Ibid., p. 36.
67. Ibid., pp. 37–42.
68. Ibid., pp. 38–39.
69. Ibid., pp. 42–79.

Chapter 7

1. Al Gore, *The Assault on Reason* (New York: Penguin, 2007), pp. 245–70.
2. John Dewey, *The Public and Its Problems* (Athens, OH: Swallow Press—Ohio University Press, 1991; orig. ed. 1927), pp. 143–84.
3. Julie Reuben, *The Making of the Modern University* (Chicago: University of Chicago Press, 1996).
4. For example, I teach a general education course called "Globalization in Historical Perspective" that includes examinations of the concepts of globalization and modernization in the social sciences, discusses economic theories of markets and international trade and theories of the state, and uses these concepts and theories to both explore the history of the global economy and understand contemporary debates in the press.
5. See, for example, an account of what happened at the University of Texas at Austin when, in the 1980s, an attempt was made to run the freshman writing course with more or less the focus that is suggested here: Linda Brodkey, *Writing Permitted in Designated Areas Only* (Minneapolis: University of Minnesota Press, 1996), pp. 181–92.

6. On this, see Jake Halpern, "Too Hot to Handle," *Boston Globe Magazine,* November 12, 2006, pp. 29–33, 43.
7. John Nichols and Robert W. McChesney, "The Death and Life of Great American Newspapers," *Nation,* April 6, 2009: 11–20. The authors of this article have developed their proposal for direct subsidy more thoroughly in a recent book: Robert McChesney and John Nichols, *The Death and Life of American Journalism: The Media Revolution That Will Begin the World Again* (New York: Nation Books, 2010). The argument seems convincing to me, but neither the book nor the article gives me much confidence that direct subsidization is likely to be politically feasible.

Index

Note: Page numbers with 'n' in the index refer to notes in the text.

abolition, slavery, 62, 66, 69, 100
academic freedom, 37, 38, 133, 172–3
academic ideology, 34–43, 48, 55–6, 81, 92, 119, 147, 161–75, 185
academic self-referentiality, 38–9
Adams, Charles Francis, Jr., 67–8, 69, 77, 102–3
Adams, Charles Kendall, 101
Adams, Henry, 64, 65, 67–8, 69, 75, 77, 101–4, 105, 107, 111, 150
Addams, Jane, 139
Addison, Joseph, 51
Adler, Mortimer, 168
advertising, 6, 17, 18, 21
affirmative action, 44–6, 219n
Agassiz, Alexander, 101
agricultural colleges, 123, 128
Alcott, Louisa May, 138
American Anthropological Association, 75
American Economic Association, 75
American Historical Association (AHA), 75–6, 130–1
American Political Science Association, 70, 75
American Renaissance, 64
American Social Science Association, 71, 73–5
American Sociological Society, 75
Anti-Corn Law League, 55
Anti-Imperialist League, 76
antislavery societies, 55

Arnold, Matthew, 78
Arsenault, Raymond, 144–5
associations, voluntary, 52, 55–6, 70, 73–7, 83, 97, 106, 130–1, 133, 157–8
Atlantic Monthly, 7, 12, 15, 23, 64, 68, 69, 70, 72, 81, 101, 104, 106, 131, 138, 141
Atlantic public sphere, 7, 55, 58, 60–5, 70, 72, 73–4, 85, 95, 181
Austria, 52

Bagheot, Walter, 69
Balliol College, 97
Bender, Thomas, 74, 99
Bennett, James Gordon, 64, 73
Berlin, University of, 85, 88, 89, 90–1, 92–3
"Berlin" or "German" university model, 92, 109–10, 111–12, 113–14, 116, 122–23, 125, 210n
blogs and blogging, 12, 14, 17, 159, 179–80, 181, 192
Boston Globe, 16
British Broadcasting Corporation (BBC), 19
British Independent Television, 19
Brookings Institution, 12
Bryant, William Cullen, 61
Bryce, James, Viscount Bryce, 68, 69–70, 77
Bush administration (2001–2009), 13, 15, 20, 197n

Cain, James M., 151
California, 27, 199–200n
California, University of, 47, 69, 117, 121, 161
Cambridge, Massachusetts, 68, 101, 102, 138
Cambridge University, 95, 96
Carnegie Foundation, 131
Cather, Willa, 139–40
Century, 68
Channing, William Ellery, 60
Chicago Tribune, 16
Chicago, University of, 156, 167
China, 3, 21
citizenship, general education for, 186–8
City College of New York, 146
City University of New York, 146
civility, 57, 107–8, 132–3, 170
Civil Rights movement, 142–5
civil service, 66–7, 73, 76, 78, 86–7, 88–9, 90, 91, 92, 94–5, 95–9, 105
civil society, 5–6, 100
Civil War, American, 3, 60, 62, 64, 65–7, 74, 79, 87, 99, 120, 139, 152
"Civil War generation", 65–8, 71, 79, 80, 99, 110, 121, 125
clubs, 6, 51, 52, 54, 74, 97, 133
coffeehouses, 6, 51–2
Cold War, 172
colleges, nineteenth-century American, 95, 100, 103, 108–9, 111–12, 113–14, 120–1, 122, 134
Columbia University, 156, 211n
Communism, 21, 173–4
community colleges, 27
Conant, James Bryant, 172–5
Conservative (or Tory) Party, British, 53
Cooper, James Fennimore, 60
Cooper Union, 122–3
core public sphere, 2–3, 6–7, 52–60, 110
  African Americans and, 83–4, 139, 142–5
  American, 7–24, 49, 65–84, 99–108, 118–20, 129–47, 177–93
  occlusion of, in twentieth century, 149–75
  women in, 58, 70, 84, 115, 137–42
  see also public sphere
Cornell University, 111, 112, 120
corporatization, 30–2, 45, 135
crisis
  of core public sphere, 2–3, 13–24, 147, 175, 177–93
  of public discourse, 1–2, 171, 193
  of public higher education, 2, 25–49, 147, 164, 174, 175, 177, 182–93
Crisis, 143
criticism, 9, 107–8, 171, 178, 179, 186–7
Croly, Herbert, 150
Cuba, 76
Curtis, George William, 68

Dartmouth College, 109
democracy, 20–1, 22, 23, 26, 49, 61, 63, 70, 84, 87, 98, 121–6, 127–9, 134–47, 152–6, 156–61, 162–3, 166, 169–70, 175, 177, 179–80, 183, 191
Democratic Party, 71
Dewey, John, 149, 156–61, 193
  analysis of the public, 156–61
Dickens, Charles, 58
Disraeli, Benjamin, 53, 55
dissertation, 91, 93
diversity, 135–47
Doonesbury, 12
Du Bois, W. E. B., 83–4, 143, 144

"economic-technological" discourse in higher education, 31–4, 35, 46, 48
Economist, 69
Edinburgh Review, 53, 55, 59, 61
Edison, Thomas, 71
elective courses, 112–14
Eliot, Charles W., 68, 69, 101–3, 110, 111, 112–13, 120
Eliot, George, 58
Ely, Richard T., 106, 118, 205n
Emerson, Ralph Waldo, 60

Enlightenment, 51, 85–8, 92
expertise, 75–6, 78, 79, 97, 98, 105, 115–20, 124, 129–31, 152–6, 158–60, 161, 164, 173–4, 179
expository writing, 124, 188–9, 191

Fabian Society, 98–9
Fascism, 21
Federalists, 61
Fichte, Johann Gottlieb, 89, 93, 94
Fish, Stanley, 201n
Fiske, John, 101, 113, 115
Fisk University, 144, 145
Flexner, Abraham, 39, 160–67, 168, 169, 173, 174, 205n
Flexner Report, 115, 129–30, 160, 161
Ford Foundation, 131
Fourth Estate, 56–7, 81
France, 6, 51, 52, 54, 55, 63, 64, 66, 76, 88, 114, 150
Frederick William III, King of Prussia, 88
freedom riders, 144–5
freedom of speech, 132–3, 172–3
Freytag, Gustav, 58

Galaxy, 68, 110–15, 115–16
Garrison, William Lloyd, 71
general education, 113, 124, 169–70, 186–9
Germany, 51, 66, 67, 71–2, 73, 75, 81, 85–95, 97, 103–4, 106, 111–12, 113–14, 116, 118, 119, 165, 172
Gilman, Daniel Coit, 68, 69, 101, 106, 110, 117, 119, 120, 165, 205n
Godkin, Edwin Lawrence, 68–73, 77, 105, 106, 150, 163, 205n, 210n
Gore, Al, 1, 17, 177
Grant administration, 66
"Great Books", 168, 169–70
Great Britain, 6, 7, 51, 52, 54, 55, 59, 60, 61, 62, 63, 64, 67, 69, 70, 76, 77, 95–9, 100, 107, 114, 119, 150
"Great Community", 159–60
"Great Society", 152–3, 154, 159–60
Greeley, Horace, 64
Guizot, François, 57

Gurney, Ephraim Whitman, 210n
Gymnasium, 89–90

Habermas, Jürgen, 5–6, 10–11, 51–3, 54, 56, 58, 60, 83
Halberstam, David, 145
Harper's Magazine, 12, 15, 23, 64, 68, 72, 106, 152
Harvard Crimson, 81
Harvard Law School, 104
Harvard University, 12, 67, 76, 101–5, 106, 107, 109, 111, 112–13, 115, 120, 121, 144, 150, 172
Hawthorne, Nathaniel, 60
Haymarket riot, 118
Hearst, William Randolph, 73
higher education, 7, 22–4, 161–75, 182
  African Americans and, 139, 142–5, 165
  and class, 138–42
  consensus model of American, 120–6, 128
  in England, 95–9
  and gender, 114–15, 119, 136–42
  in Germany, 88–95
  graduate, 80–1, 104, 114–15, 119, 163, 189
  post-Civil War change in American, 67, 72, 79, 81–2, 92, 93–4, 99–120, 189
  professional, 80–1, 90–1, 92–3, 111–12, 115, 119, 129–30, 163–4, 165
  public, 23, 25–49, 119–20, 120–6, 127–9, 133–47, 152, 173, 182–93; economic justifications for, 29–34, 175, 183; social-democratic function of, 43–8, 184
history, 54, 186–7, 189
  as discipline and profession, 75–6, 102–4, 111, 130–1
  as discursive practice, 9, 57–8, 102–4, 111
Holmes, Oliver Wendell, Jr., 65
House, Edward M., Colonel, 150
Howard University, 143, 144

Howells, William Deane, 115
Humboldt, Wilhelm von, 89–93, 96, 114, 165
Hurricane Katrina, 197n
Hutchins Report, 171–2
Hutchins, Robert M., 167–72, 174

Illinois, 122, 139
Illustrated London News, 59
immigrants and immigration, 66, 141, 145–6
imperialism, 76, 83, 97
Indian Civil Service, 97
industrial university concept, 122–3, 124
"Inquiry, The", 150
intelligentsia, 53, 161
internet, 16–17, 159–60, 177–80, 180–1, 182, 191–2
  publishing, 159, 190, 191–2
Iowa, University of, 121
Iraq, invasion of, 13–14
Iraq War, 20, 197n
Italy, 52
Ivy League, 95

James, William, 68, 101, 107
Jefferson, Thomas, 61, 100, 106, 122–3
Jewett, Sarah Orne, 138
Jewish Renaissance, 146
Johns Hopkins University, 69, 101, 106, 107, 120, 156
journalism, 9–10, 53–4, 56–7, 60–1, 66–7, 77–8, 80–1, 102–3, 135, 150–1, 161, 163, 164, 165, 171–2, 181, 189, 190–1
journals, academic, 105–6
Jowett, Benjamin, 96–9, 105

Kant, Immanuel, 85–8, 89, 91, 94
Kennan, George, 105
Kerr, Clark, 161–2, 164

League of Nations, 150
lecture, as teaching method, 92, 104, 110, 125

legal profession, 79, 81, 82, 90, 111, 129–30, 163–4
liberal arts, 29, 31, 80
Liberal (or Whig) Party, British, 53, 70
liberalism, 20–1, 71, 72, 85–8, 88–9, 94, 118, 134, 150, 154, 157
libraries, 106, 132, 192
Lippmann, Walter, 149–56, 157, 158, 159, 160, 161, 164, 167, 171–2
  analysis of public opinion and the public, 152–6
  career, 149–52
  Phantom Public, 152, 155–6
  Public Opinion, 150, 152, 153–6
Living Age, 61
Locke, John, 5
Lodge, Henry Cabot, 68, 104
London School of Economics, 98
Los Angeles Times, 16
Lowell, A. Lawrence, 76
Lowell, James Russell, 115
Lyotard, Jean-François, 47

Macaulay, Thomas Babington, 53, 54, 55, 57, 64, 67, 77, 97
magazines and periodicals, 6, 8, 12, 15, 51–2, 53, 54, 55, 56, 64, 67–8, 68–73, 76, 82, 99, 100, 102, 104–6, 108, 131, 133, 135, 138, 139, 140, 141, 143, 150–2, 158, 171–2, 174, 178, 179
  online, 12, 15, 178
Maine, 138
major concentrations in undergraduate curriculum, 113, 125, 189
Manhattan Project, 172
Marx, Karl, 55
Massachusetts, 146
McCarthy, Joseph, 20, 141, 172
McClure's Magazine, 73, 140
medical profession, 79, 81, 90, 111, 129–30, 163–4
meritocracy, 44–5, 98–9, 124, 134, 146, 174, 180
Michigan, 121, 140
Michigan College of Mining and Technology, 140

Michigan State University, 173
Michigan, University of, 111, 112, 120–21, 124, 140, 156, 210n
mining schools, 123, 140
modernity, 11
Morehouse College, 143, 144
Morrill Act, 99, 122, 124
Mugwumps, 70

Napoleonic Wars, 52, 88, 94
Nashville, 145
Nation, 7, 10, 12, 15, 24, 68–73, 74, 79, 82, 104, 106, 152, 197n, 198n
National Civil Service League, 76
National Enquirer, 12
National Review, 10, 12, 15, 24, 198n
National University idea, 122–3
nation-state, 52–3, 62, 64, 65–6, 99, 110, 152–3, 157–8
Nazism, 21
Nebraska, University of, 140
Nemec, Mark, 99
Netherlands, 76
Nevins, Allan, 99, 151
New Englander, 106
Newfield, Christopher, 45–8, 199–200n
"new journalism", 21
New Republic, 15, 150, 152, 160–1, 197n, 218n
newspapers, 6, 8, 12, 16–17, 18, 54, 55, 60–1, 62, 63, 64, 72–3, 80–1, 99–101, 105, 135, 141, 151–2, 171–2, 174, 178, 179, 181, 189, 190–91, 192, 218n
Newsweek, 12, 23, 151, 152
New York City, 62, 64, 74, 99, 122, 146, 149
New York Evening Post, 12, 59, 61, 64, 71, 72, 135
New York Herald, 64
New York Herald-Tribune, 151
New York Telegraph, 151
New York Times, 8, 12, 16, 64, 72, 81, 197n
New York Tribune, 64
New York World, 135, 151

normal schools, 123, 128
North American Review, 7, 61, 67, 69, 70, 102, 106, 131
North Carolina, University of, 121
Norton, Charles Eliot, 68, 69
novels, 53, 58, 104, 138

Obama administration, 27
objectivity, 8–9, 10, 36–7, 56–7, 116–19, 154–5, 163–5, 167, 171–2, 181, 188, 189
Olmsted, Frederick Law, 68, 69
Oregon, University of, 72
Oxford University, 95, 96, 97, 161

Peirce, C.S., 101
Pendleton Act, 78
People Magazine, 23
Philippines, 76
Plato, 96, 97
political parties, 52, 53, 54, 97–8, 135, 157–8
Populism, 83, 100, 122–4, 126, 134
Porter, Noah, 101, 111, 115
postmodernism, 36, 47, 48
presses, university, 107, 132, 190, 192
professionalism and professionalization, 9–10, 38–9, 57, 66–7, 77–82, 115–20, 128, 129–31, 142, 161
Progressivism, 77, 83, 99, 117, 123, 134, 163
Prussia, 52, 85, 86, 105
    higher education reform in, 85, 88–95
Public Broadcasting, 8, 12, 14, 18–19, 141, 182
public discourse, pre-Civil War American, 60–5
public education, 22
public opinion, 6, 10, 76, 152–6, 158–9
public sphere, 49, 51–3, 87, 88–95, 116–19, 129–47, 160–1, 170, 173–5
    "bracketing" in, 56, 75, 163, 170
    "broad" public sphere, 12, 58–60, 72–3, 102–3, 134–6, 138, 151–2, 166–7, 174, 181–2, 192

public sphere (Continued)
  Habermas's conception of, 5–6,
    10–11, 51–3
  women in, 58, 70, 138–42
  see also core public sphere
Pulitzer, Joseph, 73, 135, 151
Pulitzer, Ralph, 151
Pulitzer prizes, 81
"pure" and "applied" inquiry, 91–2, 163, 165, 185
Putnam's Magazine, 64, 68

Quarterly Review, 53, 55

racism and segregation, 20, 142–5
radicalism, 83, 94, 118, 123, 126, 160
railroads, 67, 71–2
Reconstruction, 68, 100
reform, 55, 76–7, 81, 88–95, 100
  civil service, 66–7, 71, 76, 78, 94–5, 100, 104–5
  university, 55, 80, 93–5, 95–9, 99–120
Republican Party, 68, 71, 118
research, as center of academic career, 92–3, 115–20
research universities, 27, 28, 36, 42, 47–8, 99, 161, 172, 174
Reuben, Julie, 99, 187–8
Rockefeller Foundation, 131, 161
Rockford College, 139
Rocky Mountain News, 16
Roosevelt, Theodore, 68, 75, 78
Royce, Josiah, 107
Russia, 52–3, 146

Salisbury, Robert Cecil, Marquess of, 53, 54
Salon.com, 12
Saturday Review, 55
science, 8, 55, 57, 80, 93, 99, 101, 103, 105, 109, 116–17, 129–30, 131, 154–6, 159, 163, 168–9, 172–4, 181, 189
Scotland, 51

Scripps-Howard newspapers, 151
secondary education, 89–90, 92, 96, 109, 110, 120–1, 124, 146, 166, 169, 173, 182
seminar, as teaching method, 92, 104, 110
September 11, 2001, 13–14
Sheehan, Cindy, 197n
socialism, 83
Social Question, 56
social science, 8–9, 55, 74–5, 161, 162–3, 168–9, 187–8
Spain, 52
Spectator, 51, 57
Stalinism, 172
Stanford University, 118
Steele, Richard, 51
Stein, Karl, Freiherr vom, 88
stereotype, 154, 182, 218n
Stern, Howard, 12
Stowe, Harriet Beecher, 58–60
Student Nonviolent Coordinating Committee, 145
subsidy, 15, 16, 18–19, 83, 190–1
suffragists, 138, 139

Tappan, Phillip Henry, 112, 210n
teaching, 31–2
  university and college, as profession, 9–10, 34–5, 81–2, 115–20, 130–1, 175, 184–6
technocracy, 98–9, 154–5
television, 8, 16, 18–19, 131, 141, 181, 182
Tennessee State University, 145
tenure, 37, 94–5, 133, 185, 186
Tilden, Samuel, 71
Time, 12, 152
Times (London), 55, 59, 98
Tocqueville, Alexis de, 63
transfer, interuniversity student, 125, 140–1
Trollope, Frances, 62–3
Tuskegee Institute, 143
tutorials, 97

Uncle Tom's Cabin, 58–9
universities and colleges, 8, 9, 11, 22–3, 57, 66–7, 86, 88–95, 96–9, 105, 107–8, 127–47, 158, 161–75, 182–93
   Catholic, 146
   curriculum, 108–15, 124, 167, 168–9, 173–4, 175, 186–9
   enrollments, 127–8
   financial condition of, 26–8
   on-line, 175
   organization, 108–15, 124–5
   public, 23, 25–49, 113, 120–6, 127–9, 133–47, 161, 182–93
   racial minorities in, 142–5, 146, 165
   research at, 30, 32–3, 36, 40–1, 90–1, 103, 105, 115–20, 130, 131, 132, 133, 163–5, 166, 168–9, 172–3, 183–4, 185–6, 191
   women in, 114–15, 136–42, 165

Vanity Fair, 150
Verein für Sozialpolitik, 73, 75
Vermont, University of, 112
Versailles, Treaty of, 150
Vietnam War, 20
Villard, Henry, 71–2, 77
Villard, Oswald Garrison, 71
Virginia, University of, 100, 106, 112, 120, 122–3

Wall Street Journal, 12
Washington, Booker T., 143
Washington, D.C., 62, 102
Washington Post, 151, 197n
Webb, Beatrice, 98
Webb, Sidney, 98
Wells, Ida B., 139
Westminster Review, 55
Wharton, Edith, 138
Wikipedia, 178–9, 192
Wilson, Woodrow, 150
Wilson administration, 20, 150
Wisconsin, University of, 101, 112, 118, 121, 124
women's colleges, 115, 136–7, 139
World War
   First, 20, 96, 118, 140, 150
   Second, 20

Yale University, 101, 106, 109, 111

Zola, Emile, 58

GPSR Compliance

The European Union's (EU) General Product Safety Regulation (GPSR) is a set of rules that requires consumer products to be safe and our obligations to ensure this.

If you have any concerns about our products, you can contact us on

ProductSafety@springernature.com

In case Publisher is established outside the EU, the EU authorized representative is:

Springer Nature Customer Service Center GmbH
Europaplatz 3
69115 Heidelberg, Germany

www.ingramcontent.com/pod-product-compliance
Lightning Source LLC
LaVergne TN
LVHW011816060526
838200LV00053B/3799